SpringerBriefs in Applied Sciences and Technology

More information about this series at http://www.springer.com/series/8884

Yuri N. Toulouevski · Ilyaz Y. Zinurov

Electric Arc Furnace with Flat Bath

Achievements and Prospects

Yuri N. Toulouevski
Holland Landing
ON
Canada

Ilyaz Y. Zinurov
Chelyabinsk
Russia

ISSN 2191-530X   ISSN 2191-5318   (electronic)
SpringerBriefs in Applied Sciences and Technology
ISBN 978-3-319-15885-3   ISBN 978-3-319-15886-0   (eBook)
DOI 10.1007/978-3-319-15886-0

Library of Congress Control Number: 2015933621

Springer Cham Heidelberg New York Dordrecht London

Springer International Publishing AG Switzerland is part of Springer Science+Business Media
(www.springer.com)

Contents

Introduction

The writing of this book pursued two purposes, i.e. first, to demonstrate the necessity for extensive replacement of electrical energy with the energy of fuel in electric arc furnaces (EAFs) and second, to recommend certain engineering solutions for this problem, including those developed by the authors. At present, in steelmaking the situation is quite favourable for reducing consumption of electrical energy by means of increasing consumption of natural gas. Due to the new methods of shale gas production, its price dropped sharply, and the potential for use grew significantly. Revolutionary changes have also occurred in the technology of electrical melting. The conveyor and shaft furnaces operating with flat bath have been designed and implemented. The scrap in these furnaces is continuously charged into the liquid metal bath and melted in it. Such companies as Fuchs Technology AG, Tenova S.P.A., and Siemens VAI Metals Technologies have for the most part contributed to this very promising trend.

The new technology has fundamental advantages and great potential for increasing productivity and reducing consumption of electrical energy and electrodes. Unfortunately, until now, these possibilities have not been fully realized. The reason is that both in the conveyor and in the shaft furnaces, the flow of the off-gases was used for scrap preheating. The off-gas flow did not have the heat power required. This circumstance has caused the possibilities of the furnaces with the flat bath to mainly be exhausted and their further development sharply slowed down. It is more reasonable to use the energy of off-gases rather for production of steam of technological and energetical parameters than for scrap preheating.

The abandonment of scrap heating by off-gases and using high-power burner devices instead allows to develop in the nearest future a new type of the steel melting aggregate, i.e. fuel arc furnace FAF, based on the shaft furnace with the flat bath. This unit with capacity equal to that of the EAF can provide considerably higher productivity and reduce electric energy consumption by about half. The significance of such sharp reduction in electric energy consumption goes way beyond the economic benefits of the mini-mills. It affects the prospects of steelmaking on the whole, which increasingly depends on the ever-growing requirements of environmental protection.

At integrated plants where steel is produced in oxygen converters from hot metal, specific emissions of CO_2 into the atmosphere per ton of steel are more than three times higher than emissions of CO_2 from the EAFs operating on scrap. It is necessary to point out that in case of replacing EAF with FAF, despite the fact that the emissions of CO_2 in the FAF are higher in comparison with the EAF, the combined emissions per ton of steel at both the mini-mills and at the thermal power stations supplying these mills with electrical power are 2–3 times lower.

At present, the share of the EAFs in the global steel production is less than 50 %, although these furnaces should play the leading role due to their environmental advantages. Ever-increasing reserves of unused scrap in the USA and other developed countries are favourable for this development of electrical metallurgy. In many regions of the world, a hindering factor is the lack of electrical energy resources and the sufficiently powerful electrical power lines capable of compensating electrical noise created by EAF operation. Due to operation with the flat bath, the level of electrical noise is considerably lower in the FAF. Therefore, the replacement of EAFs with FAFs could contribute to their widespread use and a substantial improvement in the ecological characteristics of the steelmaking process in general.

Electrical energy is the most universal, highly effective, ecologically clean form of energy. It should be used sparingly and only where it is really necessary. At present, consumption of large amounts of electrical energy for scrap heating and melting in the EAF cannot be justified, and replacement of electrical energy with fuel is an extremely urgent task. The authors hope that their new book will contribute to concentration of efforts of specialists in this regard.

In this book, besides the specific problems related to the development of the FAF, consideration is also given to the wide range of issues concerning an increase in productivity and efficiency of operation of a modern EAF. This book can be useful not only for developers of new processes and equipment for EAF, but also for all specialists in the field of metallurgy as well as for students studying metallurgy.

The problems of developing FAF were discussed by the authors with many specialists at the plants, in the design bureaus and scientific research institutes, which contributed to a better understanding of this issue. The authors express their deep gratitude to all of them. The authors thank Dr. Baumann for his constant attention and support of this work as well as Irene and L. Falkovich for the thorough translation of the text from Russian into English. Special thanks go to G. Toulouevskaya for her extensive work on preparation of the book for publication.

Chapter 1
Modern EAFs and Technology of the Heat with Continuous Charging of Scrap into Flat Bath

Abstract Basic technological, energy and design innovations enabling to achieve up-to-date performances of electric arc furnaces (EAFs) are considered. These innovations such as: increases in electrical power, furnace capacity, and intensity of carbon and oxygen injection into the bath; slag foaming; single scrap charging; etc., have already exhausted their capabilities in general. Prospects of further development of EAFs are associated with the new technology of continuous charging a scrap into the bath and melting it in liquid metal. This revolutionary technology partially implemented in operating conveyor and shaft furnaces creates new possibilities for sharp increase in productivity and reduction in electrical energy consumption. Analysis of these potentialities is given.

Keywords Electric arc furnace · Increase in electrical power · Increase in capacity · Intensive oxygen bath blowing · Carbon injection · Foamed slag · Single scrap charging · Furnace operation with hot hill · Continuous charging a scrap into the bath · Scrap melting in liquid metal · Increase in productivity · Reduction in electrical energy consumption

1.1 Performances of Furnaces Operating on Scrap

The effectiveness of EAF operation is characterized by many performances. The most important of them are hourly productivity, electric energy consumption, electrode consumption, and ecological characteristics. EAFs were rapidly improving. Even 20–30 years ago the present-day performances would be considered unreachable. Due to various innovations, see Sect. 1.3, tap-to-tap time of the best, most successfully operating 110–130-t furnaces have been reduced to 30–35 min. Over the last 40 years hourly productivity has increased six-fold up to 220–240 t/h. Without this sharp increase in productivity, an arc furnace would not become that steel melting aggregate which, along with oxygen converters, is shaping the steelmaking all over the world today. Electric energy consumption has

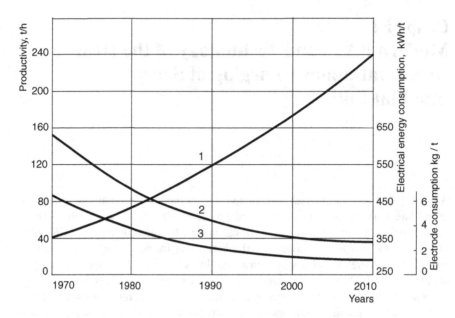

Fig. 1.1 Improvement in the 120-t EAF performances. *1* productivity, t/h. *2* electrical energy consumption, kWh/t. *3* electrode consumption, kg/t

been reduced to 340–360 kWh/t of liquid steel. Electrode consumption has been reduced approximately six-fold, Fig. 1.1. It can be expected that the majority of furnaces will come close to such performances in the nearest future.

1.2 Use of Hot Metal

Recently, in EAFs a certain quantity of hot metal is used instead of a portion of scrap. This technology has gained some spread at integrated plants where hot metal is available in excess. An optimal content of hot metal in a charge does not generally exceed 40 %. The use of hot metal increases EAF's productivity and sharply reduces electrical energy consumption since hot metal introduces a lot of physical and chemical heat into the bath. However, even with this group of furnaces the use of hot metal is rather a temporary solution than promising direction.

It is obvious that an EAF is far worse suited for hot metal processing compared with oxygen converters. In addition, EAFs dramatically deteriorate their environmental characteristics when using hot metal. Emissions CO_2 into atmosphere from EAFs, where the charge comprises of 50 % scrap and 50 % hot metal, amount to 1720 kg/t of steel, meanwhile those from furnaces operating with 100 % scrap are only 580 kg/t.

Scrap is the main raw material used in electric arc furnaces both at present and in the very long run and the reserves of scrap are great and constantly growing all over the world. This statement accords with the common long term strategies as per which demand on source materials should be satisfied to the largest possible extent due to material recycling by means of recyclability of wear and tear products and waste materials [1]. Therefore, the technology of the use of hot metal in EAFs, as well as that of reduced iron, is not examined here. Reduced iron and hot metal will be likely used only in some cases and, in the sequel, in order to reduce a content of copper and nickel in a charge when producing special steel grades.

1.3 The Most Important Technology and Energy Innovations Allowing to Achieve Current EAF's Performances

1.3.1 Out-of-Furnace Metal Processing at Ladles

At present, in EAFs as well as in oxygen converters a semi-product with given temperature and carbon content is usually melted. This metal is reduced to a final composition, cleaned of dissolved gases and nonmetallic inclusions, and heated up to an optimal temperature for casting conditions by means of ladle-furnaces, vacuum degassers, and other units of secondary metallurgy. Thereat, all process operations, which provide for steel quality required and impart special properties to it, are taken from a furnace. There is the only exception for operations of decarburization and dephosphorization of the melt, which are carried out as before in an EAF. In this manner any steel grades can be practically produced. Polluting a scrap with copper, nickel, and other residual impurities, which cannot be removed in the course of the heat, is the only obstacle for making certain steel grades in electric arc furnaces. Permissible content of these impurities for some special steels is sharply restricted. This obstacle is eliminated by a thorough preparing a scrap as well as by partial replacing it with hot metal or pig iron or with products of the direct iron reduction. Current performances of electric steelmaking could not be achieved without out-of-furnace metal processing.

1.3.2 Increasing Power and Furnace Capacity

In the past, it was an increase in the power of furnace transformers that has resulted in drastic shortening tap-to-tap time and increasing EAF's hourly productivity which required introducing a number of other essential innovations. As is known, replacing the great mass of wall and roof lining with water cooled panels, shortening power-off operation times, fundamental changing electrical mode of the heat, etc. are such innovations without which the high powers could not be successfully implemented. The electric power of EAFs increased continually. At present, that achieves 1.0–1.2 MW/t for furnaces of medium-capacity.

The maximum productivity is achieved by a simultaneous increase in the electric power and EAF's capacity. Unlike oxygen convertors, where the productivity approaches to their limit, potentialities of electric arc furnaces are still substantial. Particularly, recent development of EAFs of 300–340 t capacity with the hourly productivity exceeding 350 t/h and annual output close to 3 million of tons evidences it.

Electric arc furnaces are mostly intended to be installed at mini-mills where they determine productivity of the entire plant. Increasing output of mini-mills to one-and-a-half-two million of tons per year or even more had decisive effect on the maximum productivity level of EAF.

It is most reasonable to equip steelmaking shops of mini-mills with one furnace. Such an organization of production allows minimizing manpower and operating costs in general. If the shops are equipped with a number of furnaces then under conditions of extremely high pace of operation it is impossible to avoid some organizational delays. Any disruption of the preset production pace at one of the furnaces adversely affects other furnaces thus reducing significantly the shop productivity and that of the plant as a whole. Therefore, preference is given to the shops equipped with one furnace.

Currently, in the cases when providing the annual mini-mills output exceeding 2.0 million tons is required an installation of one 300–400-t EAF instead of several EAFs of lesser capacity is preferred. Performances of such furnaces are illustrated by the following examples. At the mini-mills in Gebze, Turkey, an EAF of 420 t capacity with tapping weight of 323 tons operates instead of four 55-t EAFs. Transformer power is 240 MVA with overload of 20 %. During the melting the active power of arcs amounts to 205 MW at the voltage of 1600 V. Charging a scrap is carried out by two baskets. The productivity exceeds 370 t/h at the tap-to-tap time of 52 min. Electrical energy consumption amounts to 326 kWh/t, electrode diameter is equal to 810 mm [2]. At the mini-mills in Iskenderun, Turkey, an EAF of 300 t capacity with tapping weight of 250 tons and transformer power of 300 MVA (1200 kWh/t) has been installed. Charging a scrap is carried out by two baskets. The productivity is 320 t/h or 2.4 million t/y [3].

1.3.3 Optimal Electrical Mode of the Heat

Electrical mode is the program of parametric changes of the furnace circuit such as current and voltage, arc power, etc. during the heat. The structure of furnace transformers enables to step these parameters over a wide range, whereas current and voltage may also vary at the constant maximum actual power. Switching over the transformer voltage steps on-load is performed either automatically or by operator's command.

Since the advent of high power furnaces their electrical modes were established based on the following common principle. During melting a solid charge while sidewalls are shielded against arc direct radiation by scrap, the maximal transformer power and long arcs, i.e., high voltage at reduced currents, are used. As liquid bath is formed the arcs are gradually shortened by voltage reduction and increase in current. At the final stages, the heat was carried out at reduced power with maximal currents which provided maximum immersion of the short arcs into the bath, the highest heat absorption by metal and minimal heat losses to water cooling the sidewall panels. Earlier it was assumed that basic principles of such an electrical mode are not subject to any revision [4].

However, these principles have undergone radical changes under the impact of general realization of the slag foaming technology, Sect. 1.3.4. Rather than shorten length of arcs for immersion them into the melt through the increased current and pressure force of arcs onto the liquid bath surface, the melt level is raised and long arcs are covered by the foamed slag. This possibility led to the development of new principles of the optimal electrical mode of the heat.

Electrical power can be increased due to the step-up of both voltage and current since the power is proportional to the product of these values. Both these directions are tightly associated with the problem of graphitized electrodes which is one of bottlenecks limiting further increase in electrical power. Electrode current density, which is severely limited to avoid the sharp reduction in electrode durability, goes up with an increase in current. At present, electrodes of UHP furnaces operate at the current densities of $25-35$ A/sm^2. Further increase in the current density requires substantial improvement in qualitative characteristics of electrodes such as: conductance, density, strength, etc., resulting in heavy complicating of manufacturing technology and dramatic increase in electrodes cost.

In order to increase electrode current-carrying capacity there is a way consisting in increase in electrode diameter. Alongside with wide-spread electrodes of diameter 610 those 710 and 810 mm are used [2]. However, increasing electrode diameter is attended with overcoming difficulties no less significant than the growth of current-carrying capacity and results in sharp rise in electrode cost. Today, the possibilities of both directions are basically exhausted. On the contrary, raising electrical power of EAFs due to increasing maximum secondary voltage of transformer still has reserves. Previously, this voltage did not exceed 1000 V. At present, furnaces operate successfully at voltage of 1350–1600 V.

Voltage increase has another considerable advantage compared with current increase. The former is not accompanied by the increase in electrode consumption which is approximately half-determined by current density and rises proportionally to the growth of this value. However, the opportunities of further voltage increase are also limited by a number of factors including the dander of spark-over in dust-loaded gaseous gap between electrode holders over the roof as well as between the electrodes and the furnace roof.

Method of foamed slag allows to outline the optimum electrical mode of a modern EAF as operation at constant electrical parameters during the entire heat, i.e. at maximum active arc power at maximum voltage and minimum current for the given power value. Such a mode enables to obtain the maximum furnace productivity at minimum steelmaking costs. The closest to the implementation of the optimum mode were furnaces operating with continuous scrap charging into the liquid bath, Sect. 1.6. However, this mode can be also implemented successfully on the furnaces operating with a single scrap charge. Only at the very beginning of the heat, while arcing takes place above the scrap pile, in order to the avoid furnace roof damage, slight arc shortening, achieved by minor voltage lowering and current increase, could be required.

1.3.4 Foamed Slag Method

One of the most significant results of implementation of new engineering devices for joint bath blowing with oxygen and carbon powder was the possibility of reliable slag foaming and maintaining slag foam thickness at a layer assuring complete immersion of electric arcs in slag. Such a technology increased heat transfer from arcs to the melt up to the maximum possible level and provided a number of other advantages.

Slag foaming mechanism, during concurrent blowing oxygen and carbon into the bath, is as follows. Oxygen oxidizes carbon contained in the bath and dissolved in it, by reaction $C + 0.5 O_2 \rightarrow CO$. A certain portion of oxygen is consumed for iron oxidation with formation of FeO. Carbon injected into the bath is dissolved there and reduces iron oxides by reaction $FeO + C \rightarrow Fe + CO$. Thus, both reactions running concurrently generate small bubbles of CO gas which float upwards foaming the slag. If proper correlation of oxygen and carbon consumptions is assured FeO reduction by carbon allows to use oxygen in large volumes without any fear of yield drop.

Carbon consumption for creating a foamed slag layer of required thickness depends on the arc length and amounts to approximately 6–10 kg/t in modern furnaces. However, in order to increase furnace productivity in practice this consumption is approximately doubled which also allows increasing oxygen consumption significantly. It should be stressed that excessive thickness of foamed slag layer reduces the productivity and other basic parameters of furnace performance. Therefore, it is very important to develop reliable means of controlling the slag foam level.

Consumption of carbon powder depends on its quality as well as on injection method and slag composition. The quality of this expensive material is mainly determined by the content of carbon in it and by the ability to dissolve quickly in the melt. Coke powder contains on average only as much as 80 % of carbon. Carbon content in graphite powder is considerably higher. Besides, graphite dissolves much quicker.

To achieve as complete absorption of injected carbon as possible, it is quite important to distribute points of injection around the bath perimeter; the distance from the injectors to liquid metal level is also of great significance. Of all practiced injection methods, the best carbon assimilation, close to 100 %, is achieved when carbon is injected directly into the slag near slag-metal interface. The worst option is to inject carbon powder from above onto the surface of slag. In that case, a significant portion of the material is removed out of the furnace by off-gas flow.

The selection of injector locations is of great importance for the efficiency of carbon injection. It is not recommended to combine carbon and oxygen injection points. When such combining occurs, oxygen and carbon jets come into direct contact with each other, so carbon burns out partly prior to being dissolved in the melt. This portion of carbon is uselessly lost both for slag foaming process and for iron oxides reduction. Therefore, the oxygen and carbon injectors should be located at a certain distance from one another. Carbon injection into the bath zone in front of the slag door is inexpedient since foamed slag tends to flow over the furnace sill, and carbon is partly lost for the process. In this regard, the best results are provided by carbon injection into oriel zone and the bath zone adjacent to it since foamed slag moves towards the arcs and covers them.

Foaming ability and stability of formed foam depend greatly on the physical properties of slag, such as its viscosity, density, surface tension, and the concentration of undissolved solid particles. All these properties are determined by slag composition and its temperature. In basic slags, foaming ability increases as SiO_2 concentration grows. In the modern EAFs the duration of liquid bath existence is rather short-term, and there is not enough time for slag to be completely formed. The slag is rather non-homogeneous and contains great quantities of undissolved particles of lime and other small-size particles. That contributes to better and easier foaming and excludes any necessity of increasing SiO_2 concentration in the slag significantly since it reduces basicity of slag compared with the ordinary level. Foaming is also facilitated by injecting dolomite and lime powder into the slag, as well as by adding coke to slag at an early melting stage.

In order to keep a thick layer of foamed slag without its flowing over a sill, a slag door is shut. At the same time, air infiltration into a freeboard is eliminated. The mode of operation with a shut slag door is in common use on the modern furnaces. This mode of operation, however, leads to forming of sizeable metal-slag buildups on a long furnace sill in front of the closed slag door. These buildups hinder monitoring of temperature and metal composition during the heat and cause difficulties in furnace maintenance. During the heat it is required to open the door in order to purge the sill for slagging-off. For this purpose various pushers and other special mechanisms, which occupy a lot of room at operating platform, are used.

To eliminate these shortcomings, V. Shver (PTI Inc. Company, the USA) has developed an original design of a slag door, so called SwingDoor™, Fig. 1.2. The key feature of this water-cooled door is that it moves around its upper horizontal axis rather than up and down as it typically takes place. In a shut position, this door becomes as an extension of the wall panels, which precludes freezing of slag on the sill. The sill is free from buildup along its full length, thus all the complications associated with buildup removal are eliminated. For slagging-off the slag

Fig. 1.2 SwingDoor™

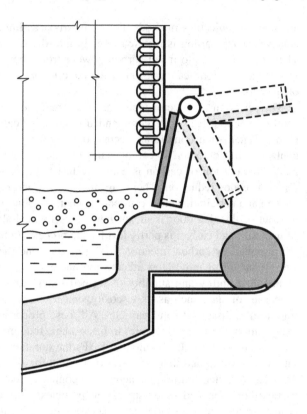

door is turned through angle of about 40°. In so doing, no air infiltration into a freeboard is allowed. The slag door is opened in full solely in case of shutdowns of the furnace in order to examine and repair refractory of bottom and water-cooled panels.

1.3.5 Furnace Operation with Hot Heel

At present, a steelmaking method is widely used where about 15–20 % of metal and certain amount of slag are left at the furnace bottom after each tapping. The rest of slag is removed from the furnace over the sill. If steel-tapping hole is made according to the modern requirements, Sect. 1.4.2, such a method allows to pour practically slag-free metal into the ladle. That provides ferroalloys savings and facilitates performing subsequent operations of secondary metallurgy. Availability of hot heel betters conditions of dephosphorization. Yet, the main advantage of EAF operation with hot heel is energy efficiency.

In high power furnaces, boring-in scrap pile occurs so quickly that melt layer is not deep enough when electrodes reach closely to the bottom. There is a danger of damaging bottom refractory by powerful arcs. This factor restricts increasing electric power of the furnaces. Presence of hot heel eliminates the said limitation and allows increasing electrical power with the aim of further productivity increase.

Operation with hot heel extends the capabilities of effective use of oxygen for blowing the bath which also promotes growth of productivity. Presence of hot heel allows starting oxygen blowing almost immediately after scrap charging. When blowing hot heel in the presence of carbon charged with scrap, slag is foamed and the arcs immerse into the melt, thus increasing their efficiency. Carbon monoxide CO, escaping from the hot heel, combusts and heats the scrap layer when passing through the scrap thus accelerating settling and melting down of charged metal-charge. Oxygen blowing of relatively cold scrap pile, when started too early without hot heel, is ineffective. Although such blowing accelerates the heat but this is achieved due to intense iron oxidation which leads to unjustified yield drop. At the same time, electrodes oxidation and consumption increase as well. Operating with hot heel provides maximum advantages at EAFs with continuous scrap charging into the bath. At these furnaces, a hot heel weight is significantly increased which allows increasing a charging rate and shortening tap-to-tap time.

Maintaining the mass of metal, left in the furnace, at a relatively constant level close to the optimum is a necessary condition of substantially complete and stable using the advantages provided for by operation with hot heel. Significant fluctuations of hot heel size from one heat to another significantly reduce the efficiency of this important process element. Visual control is insufficient. It is necessary to use engineering tools to enable easier operator's control of hot heel. For this purpose, furnaces are equipped by sensors which allow controlling a furnace weight varying during the heat and consequently a hot heel weight.

1.3.6 Intensive Use of Oxygen, Carbon and Chemical Heat

Alongside with electrical power increase, this innovation has played an exceptionally great part in increasing EAF productivity and reducing electrical energy consumption due to the fact that the bath absorbs a large quantity of chemical heat released when oxidizing carbon, iron and its alloys as Si, Mn, etc. by oxygen. Methods of injecting oxygen into the freeboard and liquid bath were improved continuously.

At first, oxygen was used in EAFs in rather limited amounts, mainly for cutting scrap and bath decarburization. Oxygen was injected into the furnace manually through a slag door using steel pipes. Later on, this operation was mechanized completely. To introduce oxygen various manipulators started to be applied, and not only consumable pipes but also water-cooled lances were used which were inserted through the openings in the roof and sidewalls of the furnace. Sidewall oxy-gas burners of 3–3.5 MW were also widely used in electric arc furnaces. All that contributed to increase in oxygen consumption. It should be noted, that in the 80 s of last century, oxygen consumption in the bath did not exceed 10–15 m^3/t of steel.

Further sharp increase in oxygen consumption, especially for the bath blowing, is inseparably linked to use of carbon powder injected into the bath concurrently with oxygen. The impressive results achieved by expanded oxygen use would have never been obtained without carbon injection. As the intensity of oxygen blowing of the bath increased, the amount of oxidized iron increases inevitably. Carbon reduces iron oxides thus preventing unacceptable drop of yield. Besides, injected carbon causes slag foaming. Immersion of arc in foamy slag assures sharp increase in electrical energy efficiency, Sect. 1.3.4.

In modern high power furnaces operating on scrap, an average oxygen consumption is approximately 40 m^3/t, not infrequently, it is as high as 50 m^3/t. Oxygen consumption increases, in some cases, to 70 m^3/t in furnaces where oxygen is used also for post-combustion of carbon monoxide (CO) in the freeboard. Consumption of carbon powder injected into the bath reaches as high as 15–17 kg/t.

Since the heat duration is very short, such specific oxygen and carbon consumption values require quite high-intensity injection. In modern furnaces, the specific intensity of oxygen blowing is usually 0.9–1.0 m^3/t per minute and it may also reach 2.5 m^3/t per minute if hot metal and reduced iron are used in large amounts; the latter value approaches the blowing intensity observed in oxygen converters.

Wide application of oxy-gas burners and oxygen in modern EAFs has sharply increased a share of chemical energy in total amount of energy consumed per heat. The share of electrical energy was reduced to approximately 50 % of total energy consumption, whereas in furnaces where hot metal and reduced iron are used it dropped even significantly lower. As far as energy aspect is concerned, such furnaces have very little in common with the furnaces of the past, where the role of other energy sources was quite insignificant compared with electric arcs. It could be stated today that the energetics of electric arc furnaces is quite close to that of oxygen converter processes based exclusively on the use of internal sources of chemical heat.

1.3.7 Single Scrap Charging

Recently, EAFs with expanded freeboard size capable of receiving all scrap of about 0.7 t/m^3 volume density charged by single basket are getting spread. Charging each basket requires roof swinging and current switching off. With short tap-to-tap time, using one scrap basket instead of two leads to a considerable increase in EAF's hourly production. However, the advantages of furnaces with single scrap charging are not limited to that.

Freeboard volume is expanded in such furnaces mainly by means of increasing its diameter and, essentially, height. Greater height of scrap pile in the furnace provides for better absorbing by scrap of the heat of hot gases, obtained when post-combusting of CO and passing upwards through scrap layer from below. The same can be said about absorbing heat from flames of oxy-gas burners installed in the lower parts of furnace sidewalls. Increasing depth of pits bored-in by arcs in scrap also increases the degree of arc heat assimilation. All this increases scrap heating temperature prior to its immersion into the melt, and accelerates melting. At the same time, electric energy consumption is decreased due to the reduction in the time when the furnace is open and loses a lot of heat. Dust-gas emission into shop atmosphere is reduced also while scrap charging. In furnaces of 300–400 t capacity, due to the freeboard height extend, the number of charges is decreased to two per heat.

However, considering the effect of increasing the furnace freeboard height on the utilization of heat in it, it should be taken into account that sidewall area is increased and, consequently, heat losses with cooling water are increased as well. To reduce these losses, measures are taken to increase the thickness of skull layer on the sidewall panels. For instance, Danieli Company uses panels consisting of two layers of pipes. The pipes of the internal (with respect to freeboard) layer are spaced apart much wider than in the external one. That facilitates formation of thicker skull layer and its better retention on the pipes [5]. As freeboard height is increased considerably, electrode stroke and their length are respectively increased as well, thus increasing the probability of electrode breaking. To prevent breaking the rigidity of arms and of the entire electrode motion system should be increased. Lateral surface area of electrodes as well as their wear due to oxidation, which comprises about 50 % of the total electrode consumption, are increased. Taking into consideration all these factors it is believed that furnaces of no more than 180 t capacity are most suitable to realize a single scrap charging.

It should be paid attention to the fact that a freeboard height required during the scrap charging and melting period comes into conflict with an optimum height after the flat bath formation when this height should be significantly shortened in order to reduce heat losses. In EAFs with a single charging this contradiction is considerably aggravated. Designs of furnaces with a variable freeboard due to use of a telescoping shell have been developed. Continuous scrap charging in furnaces with a flat bath allows making optimal freeboard dimensions identical for all periods of the heat and solving a problem to the best advantage.

1.4 Most Important Design Innovations

1.4.1 Optimization of Freeboard Dimensions

Optimization of the freeboard dimensions is one of the most important design innovations which results from great experience in designing and operating of arc furnaces gained worldwide. The typical bath dimensions of modern furnaces are given in Table 1.1. It is assumed that a weight of steel tapped into a ladle amounts to about 85 % of liquid metal in a furnace and that approximately 30 % of liquid metal is found in an oriel and about 70 % in a bath itself. In 80–120-t EAFs, a

Table 1.1 Typical
dimensions of a freeboard in
modern electric arc furnaces

Tapping weight (ton)	Depth of a bath (mm)	Bath diameter at a sill level (m)
80	1150	5000
120	1350	5600
300	1700	7300 (max)

ratio of bottom lining thickness to bath depth drops from 0.62 to 0.52 as capacity increases. As capacity is further increased, bottom lining thickness changes insignificantly and remains at a level of 800 mm.

A weight of scrap charged into a furnace is determined taking into account a yield of liquid metal which amounts to 90 % in high-capacity furnaces operating with state-of-art means of intensification and with metal-charge of relatively low quality. Using the following symbols: M_L and V_L as weight and volume of liquid metal tapping into a ladle; V_S as volume of scrap; 07 V_L as volume of main part of a bath without including a hot heel; D as diameter of a freeboard at panels level; H as freeboard height from a sill to the top of sidewalls; ρ as density of liquid metal ($\rho = 7$ t/m^3); ρ_s as volume density of scrap ($\rho_s = 0.7$ t/m^3), we obtain:

$$M_S = M_L/0.9; \quad V_S = M_L/0.9 \cdot 0.7 = 1.59\,M_L.$$

An overall freeboard volume is comprised of volume of the main part of a bath and that of the upper freeboard part: $V = 0.7M_L/\rho + 0.785D^2 \cdot H$ or $V = 0.1\,M_L + 0.785\,D^2 \cdot H$.

For a single scrap charging, a condition of equality $V = V_S$ must be satisfied.

Given that on average D/H $= 2$, then for the 80–100-t furnaces we get $D^3 = 3.11\,M_L$. This expression can be used to determine dimensions of the freeboard for the furnaces of this capacity.

Selection of both electrodes and electrodes circle diameters has great importance for EAFs' operation. Electrodes of diameter $d_E = 610$ mm have become the most widespread in the world practice. In AC EAFs they are used for the currents up to 80 kA. Electrodes of diameters 700–750 mm are used for the higher current loads, and for the currents higher than 100 kA the diameter is increased up to 810 mm. As the diameter of electrodes increases, their price and expenditures on electrodes go up significantly. A weak spot of electrodes limiting their current load is the bottom electrode nipple joint affected by the highest temperatures.

A number of factors affect selection of optimal electrodes circle diameter d_O. Introduction of electrode holders with current conducting arms have made possible significant reduction of this diameter, which in turn accelerates melting of scrap.

For instance, at Magnitogorsk metallurgical complex, on a modern 180-t EAF operating at 150 MVA transformer power, secondary voltage of 1400 V, and electrodes circle diameter of 1200 mm, the time of penetration of arcs through a column of scrap is no more than 6 min, after which the arcs are burning on liquid hot heel. As this occurs, a rate of absorption of arc heat by scrap increases, since the walls of a deep cave in a column of scrap shield water-cooled panels from radiation of arcs. However, the practice shows that this increases a risk of electrodes breaking due to heavy lumps of scrap crashing down from the top part of the cave.

In modern EAFs operating at a secondary voltage of 1600 V, an unreasonable reduction of the electrodes circle diameter results in electrical breakdowns at gas inter-electrode gaps which are heavily dust-laden as well as at the gaps between elements of the electrodes holders. Moreover, concentration of high power in a small inter-electrode zone of the bath causes overheating and evaporating of liquid metal and, consequently, reduces the yield. All this shows that a tendency to decrease electrodes circle diameter d_O is justified only to a certain degree. In furnaces operating with a secondary voltage of 1600 V, the ratio d_E/d_O should be increased to approximately 2.4. For modern 80–240 ton EAFs the reasonable ratios D/d_O are in the range of 5.2–6.2.

1.4.2 Slagless Tapping System

When carrying out ladle refining of steel in ladle-furnace units, in order to effectively carry out desulfurization and reduce burn-off loss of alloying elements it was necessary to eliminate entry of slag with high content of iron oxides from EAF into a ladle. This necessity has led to developing of furnaces with bottom tapping. First furnace of this type with 90 t capacity was introduced in 1976 at the Thyssen Company plant. Then such tapping has transformed into eccentric tapping with taphole which has been moved relative to the centre of the furnace. However, electric arc furnaces with oriel tapping have become the most widespread. Diameter of oriel opening in EAFs of various capacity ranges from 150 to 220 mm. The opening is filled with a special (most often, dunite) powder.

To tap a heat, a steel ladle car is moved under a furnace which is tilted up to 5°–6°, and then a taphole gate is unlatched. Powder pours out and metal is tapped into a ladle. As metal is being tapped, a furnace is tilted to 12°–15°. A required quantity of tapped metal is determined using the readings of tension sensors installed on a steel ladle car. By the end of tapping, a swirl on the melt surface is formed which starts to draw slag into the metal stream. At this particular moment a furnace is promptly returned to the initial position leaving hot heel in a bath. Thus, availability of hot heel is a necessary condition for slagless metal tapping from a furnace. When a furnace is returned to the initial position, a taphole is closed by a gate and filled up with dunite powder through an opening in an oriel cover. If it is necessary to completely tap a metal for furnace maintenance, a furnace is tilted to 25°.

The presence of slag from a previous heat along with lime introduced into a freeboard allows to quickly produce a slag of high basicity in the liquid bath. This creates necessary conditions for dephosphorization of a metal. The intensive slagging-off over the slag door sill precludes rephosphorization during further heating of the bath.

In addition, the following advantages of the oriel tapping should be mentioned:

- Shortening metal tapping time by 1–2 min
- Tapping is carried out in a form of short concentrated stream which assures minimum heat losses and minimum saturation of metal by gases
- Decrease in tilt angle during tapping. This allows to increase an area of water-cooled sidewall panels, decrease a weight of metal structures of a furnace, and reduce mechanical loads on electrodes and masts of electrode holders
- Reduction in length of flexible cables of current-conducting arm

1.4.3 Electrode Holders with Current-Conducting Arms and Gantry-Less Design of EAFs

An important development in EAF's design is the electrode holders with current-conducting arms which increase reliability of furnace operation and reduce electrical energy consumption. Design of these holders has been developed by Fuchs Systemtechnic company and first implemented in a 36-t EAF in Spain, in 1983. In this electrode holder design, a water-cooled arm made of copper-clad steel sheet functions as a current-conductor. For EAFs of different capacities, a copper layer thickness is 4–6 mm and that of steel layer is 10–22 mm. Spring-hydraulic electrode clamping mechanism is placed inside the arms. In the furnaces operating with currents from 20 to 100 kA, electrode clamping force of 20–70 ton is provided by means of disk springs whereas electrode releasing is carried out with the help of hydraulic cylinder designed for operating pressure of 16–25 MPa. In UHP EAFs of medium and high capacities, 2 or 4 cables with cross section of 3600–5400 mm^2 in each phase are used as flexible current-conductors.

Application of electrode holders with current-conducting arm makes it possible to:

- improve rigidity of group of electrodes as well as furnace operation reliability,
- allow minimum electrodes circle diameter,
- decrease active resistance and inductive reactance and thus reduce electrical energy consumption by 3–5 %.

Among other new designs, the design of gantryless EAFs is worth mentioning. In these furnaces, the frame of peripheral area of the roof and, in some cases, the ring of its central area are connected to the carrying beam equipped with a landing ring. A piston rod interlocks with this ring and provides lifting and swinging of the roof. Usage of the gantry less design allows to decrease an overall weight of metal structures and to free some room over the roof for installation of process equipment.

1.4.4 Speeding-Up of EAF's Drives and Shortening of Power-Off Times

A pace of process operations of modern EAFs has reached such level of intensity that not only minutes but also seconds count. Charging any slag-forming additives, ferro-alloys, carburizers etc., is carried out without switching-off a furnace. The time of lifting and swinging of the roof has been shortened down to 20 s due to the use of a proportional hydraulic distributor. The speed of electrodes movements has been increased to 250–300 mm/s. The speed of tilting a furnace for tapping is 2–3°/s and that of returning a furnace is 6–8°/s.

Among other innovations assuring shortening of power-off time we should mention the craneless electrode slipping during swinging of the roof for charging scrap into furnace as well as the use of remote-controlled manipulator for sampling and taking temperature. A great attention is also given to main lifting and moving speeds of the crane serving a furnace. For example, a 240-t bridge crane used for charging scrap into furnace with the help of a charging bucket has a bridge movement speed of 75 m/min, car speed of 35 m/min, and main hook lifting speed of 10 m/min. In addition, the crane is equipped with a positioning system and is operated by remote control from an operating platform.

Slag removal has become of great importance in high-efficient EAFs. In case of the state of the art technology, an amount of slag formed in a furnace is 6–12 % of the weight of steel produced. A conventional method of slag removal using a slag pot becomes a limiting factor when the slag foaming technology is used and fits poorly into the system of modern electrical steelmaking.

In modern EAFs of more than 40 t capacity, pot-less slag removal is mainly used. In these furnaces, a subspace between the bearers of a basement (about one third of the length) is partitioned by a cross wall. A basement is bordered with steel plates or water-cooled panels. As a result, a slag corridor is being formed, and foamed slag flows by gravity over the slag door sill into this corridor. Slag removal is carried out by either wheeled autoloader or by crawler loader.

1.5 Change of Method of Scrap Charging as a Pre-requisite for Improvement of Main Performances of EAF

1.5.1 Increase in Productivity

The possibilities for further increase in productivity of EAF for the most of innovations discussed above are basically exhausted. An example of such innovation is secondary ladle metallurgy, since all of its possible operations are already being carried out outside of the furnace. Increasing of EAFs capacity has also practically reached its limit. Further increase of intensity of bath blowing leads to reduction of effectiveness of oxygen and carbon usage as well as yield reduction. A relative mass of hot heel has also reached its reasonable limit. The same relates to the secondary voltage of EAF transformers. Concerning an improvement of electric power utilization due to shortening of power-off furnace operation periods τ_{off}, there is still some potential. During τ_{off} period, the operations requiring turning off of a furnace are being carried out. These are (completely or partially) tapping, closing of taphole, electrode slipping and building-up, fettling the bottom banks, scrap charging, etc. An increase of productivity requires maximum possible shortening of τ_{off}. At the same transformer power, decrease of τ due to reduction of τ_{off} makes it possible to increase an average electrical power as well as an hourly productivity of a furnace, which grows directly proportional to reduction of τ.

One of the most effective ways of shortening τ_{off} is decreasing the number of scrap charging as well as reducing duration of all the operations related to them. For example, on one of the EAFs, duration of charging of one basket along with operations of roof opening and shutting has been reduced to 50 s [6]. Such an outstanding result was achieved due to increase of the operating speed of roof swinging mechanisms, but, mainly, because of extremely well-coordinated actions of well trained, highly-skilled personnel. Normally, charging of one basket of scrap takes about 2 min. Thus, single charging allows to shorten τ_{off} and τ by 1–2 min as compared to charging with 2 baskets.

Continuous scrap charging into liquid bath or charging of scrap by small discrete portions quickly following each other (which practically means the same) allows to completely eliminate time spent on conventional charging of scrap from the top. This way of charging does not require increasing of freeboard height which is needed for single charging process. Consequently, all the shortcomings of single charge resulting from increasing of freeboard height are overcome. Yet the advantages of continuous scrap charging extend beyond just shortening of τ_{off}. Such charging process also enables increasing of the electrical energy efficiency of EAF.

As already mentioned above, Sect. 1.3.3, electric power supplied to a furnace should be maintained at a maximum level for most of the power-on furnace operation time. Based on modern views, the optimal electrical mode of EAF is such that the electric power of a furnace is constant and corresponds to maximum power of the furnace transformer. Such a mode allows to achieve the highest hourly productivity.

Conventional charging of scrap from the top which has remained unchanged for many decades hinders implementation of the optimal electric mode. In case of such charging, the conditions of thermal operation of electric arcs change drastically over the course of the heat which requires adjusting the inlet power in the furnace. At the beginning of the process, the power of arcs burning close to the furnace roof has to be reduced because of the danger of roof damage. After forming of flat bath still containing unmolten scrap, the electric power has to be reduced as well to avoid metal overheating and damage of banks and bottom refractories. During the stage of caving-in of a column of scrap, the lumps of scrap crash down quickly. Under such conditions, arcing is unstable and the power fluctuates widely which leads to reduction of its average value. As a result, high electric power of the furnace is utilized insufficiently. Continuous scrap charging stabilizes conditions of arcing and allows to keep power introduced practically at constant level, which increases the furnace productivity as well as the electrical energy efficiency.

1.5.2 Reduction of Electrical Energy Consumption

Electric energy is ecologically clean, most useful with regard to technological aspects, but the scarcest form of energy. Currently existing in the world ratio of prices on fossil fuel energy and electric energy is artificially sustained and is nowhere near their real value.

The global electric power industry is based on electrical power production by means of thermal power stations (TPS) which use coal and natural gas and pollute the environment with CO_2 emissions. Share of hydroelectric industry is relatively small, and so-called alternative energy sources, i.e. the sun, wind, and sea waves, play quite negligible part. The situation will remain the same over the long term considering that, in spite of widespread opinion, global fossil fuel reserves are practically unlimited. Therefore, in accordance with the requirements of environmental protection, electrical energy must be used only when it is really essential. Thus, for example, in accordance with the long-term strategy for development of global power industry, persevering efforts are undertaken to convert, in the relatively near future, all motor vehicles to electric vehicles, since transportation using internal combustion engine is one of the major sources of air pollution [1]. Taking this into consideration, the usage of electric energy for scrap heating in EAF, at the very least, is unreasonable.

To estimate electrical energy consumption for scrap heating in EAF, let us refer to Table 1.2. This table will be repeatedly referred to in other chapters of the book. As per Table 1.2, the enthalpy (or, in other words, heat content) of liquid metal in case of tapping temperature of 1640 °C is 394 kWh/t. Enthalpy value of liquid metal before tapping we will call overall useful energy (or heat) consumption for the heat. Metal obtains this amount of energy during the heat from all used energy sources, i.e. electric arcs, burners, the exothermic chemical reactions of oxidation of iron and its alloys which proceed with heat energy release, etc.

The enthalpy of scrap is assumed to be equal to zero at t = 0 °C. The useful energy consumption for scrap heating to the melting point 1536 °C is equal to a difference between the enthalpies at these temperatures, or 294 kWh/t, which comprises almost 75 % of the total useful heat consumption during the process: 294/394 = 0.746. At that, 369 − 294 = 75 kWh/t, or 19 %, is utilized for melting of the heated scrap, and only 394 − 369 = 25 kWh/t, or 6 % of the total useful heat consumption, is utilized for heating of the melt to tapping temperature, Table 1.2. Hence, almost three quarters of the heat required to produce the liquid steel at the tapping temperature are utilized for heating of scrap to the melting

Table 1.2 Heat capacity c and enthalpy E of solid and liquid iron at various temperatures t

t, °C	c, Wh/kg °C	E, kWh/t	t, °C	c, Wh/kg °C	E, kWh/t
25	0.124	3.1	900	0.190	171
50		6.5	950		181
100	0.130	13.0	1000	0.190	190
150		19.8	1050		198
200	0.134	26.8	1100	0.188	207
250		34.1	1150		216
300	0.139	41.6	1200	0.187	225
350		49.5	1250		234
400	0.144	57.7	1300	0.187	243
450		66.5	1350		254
500	0.155	77.5	1400	0.190	266
550		85.0	1450		275
600	0.158	95.1	1500	0.191	287
650		106	1536 solid		294
700	0.169	119	1536 liquid	0.240	369
750		132	1600		384
800	0.181	145	1640	0.240*	394
850		158	1680		403

Notes Enthalpy of low-carbon scrap is higher than those by 2–3 %
*Average heat capacity over a range of temperatures 1536–1680 °C

point. Therefore, both productivity and efficiency of furnace operation depend first and foremost on how well the scrap heating process is set-up. At present, scrap heating is carried out mostly by electric arcs. The contribution of the burners into this process is auxiliary and relatively small.

This is the case despite the fact that no particular requirements are imposed on the scrap heating quality; it does not affect the quality of produced steel in any way. Yet, in the furnaces for the high-temperature heating of rolled billets and in the furnaces for heat treatment of rolled steel, including heat treatment in neutral atmosphere, where the quality of heating is of paramount importance, preference is given not to electrical heaters, but to fuel burners, including flame-protected radiation burners, in spite of some related serious technical difficulties. The use of electric energy along with the use of simple and reliable means of automatic control could ensure the best possible quality of heating. However, due to economic reasons, even for these purposes the burners are mostly used. Let us review the reasons why, despite the economical factors, the burners for scrap heating in EAF play a secondary part and the possibilities of wide substitution of electric energy by burners still have not been realized.

In EAFs low-power oxy-gas burners became overall spread. As a rule, unit power of such burners does not exceed of 3.5–4.0 MW. They are installed in side-wall panels at a height of about 500 mm above the bath level as well in the oriel covers and the slag doors. In the past, three sidewall burners used to be installed in EAFs in so-called cold spots between the electrodes where closely walls the scrap melting was delayed. Sidewall burners equalized a temperature profile around all the periphery of the furnace. Oriel burners eliminate cold spots at the oriel and sidewall ones do that at the slag door sill area. The latter allows an earlier taking metal sample and temperature which contributes to shortening the heat. With the low unit power such an application of burners does not make a great impact on electrical energy consumption.

Further practice has resulted in understanding the fact that to increase fuel consumption in the burners is necessary rather for intensification of the process than saving electrical energy. With continuous shortening tap-to-tap time a significant increase in the burner power was required. However, all attempts made in this direction have not given positive results. Today, the unit power of burners, due to the reasons described in detail below, is at the same level as 20–30 years ago. Therefore, in order to increase overall power of the burners the number of burners has been increased. The number of burners in the furnaces reached 6–9, and even 12, as in the Danarc Plus furnace, where they were installed in two rows with the height of the shell [5].

Since the early nineties installing the burners inside water-cooled copper boxes, which protected the burners against both high temperatures influence and damages to those when charging scrap, has been started. Thereat the burners have been combined with nozzles for the bath blowing with oxygen.

Despite the increase in the number of burners, specific consumption of natural gas in the furnaces did not grow significantly. Usually, it does not exceed 5–6 m^3/t. This is a result of the further reduction of the tap-to-tap time and, correspondingly, burners' operation time. The effectiveness of the burners did not change as well. As before, they ensure reduction in tap-to-tap time and electrical energy consumption by not more than 6–8 %.

In the vast majority of cases, natural gas is used in the burners of EAF. All conventional burners are similar in general principles, regardless of a furnace size and their location. The design of these burners provides for intense mixing of gas and oxygen firstly in the combustion chamber of the burner and then does that close to its orifice. When used for scrap heating, the burners operate with oxygen flow rate coefficient of approximately 1.05. Usually, they form a narrow high-temperature flame. Initial flame speeds are close to the sonic speed or exceed it; maximum flame temperatures reach 2700–2800 °C. When the burners are used for scrap cutting or for post-combustion of CO, the oxygen excess is increased to 2–3.

Heating of liquid bath with burners is ineffective. However, small amounts of both gas and oxygen have to be supplied to the burners to maintain the so-called pilot flame. This allows to avoid clogging of the burners with splashed metal and slag. These forced non-productive consumptions of energy-carriers noticeably worsen burners' performance indices including energy efficiency coefficient of gas η_{NG}. A pilot flame only diminishes but does not completely eliminate clogging of the burners. Periodically, they have to be purged, which poses certain difficulties considering the number of burners.

Let us review the causes hindering the increase in power of burners and their application for high-temperature scrap heating. During the operation of these burners, the direction of flame remains constant. Burner flames attack the scrap pile from the side, in the direction close to radial. The kinetic energy of the flames of conventional burners is low due to their low power. Penetrating into a layer of scrap these flames quickly lose their speed and are damped out. Therefore, their action zones are quite limited.

Since emissivity of oxy-gas flame in the gaps between the scrap lumps is low, heat from flames to scrap is transferred almost completely by convection. With convection heat transfer, the amount of heat transferred to scrap per unit time is determined by the following aerodynamic and thermal factors: the surface area of the scrap lumps surrounded by gas flow; the speed of gas flow which determines the heat-transfer coefficient α; and the average temperature difference between gases and heat-absorbing surface of the scrap. In the action zone of the burners, at high temperatures of oxy-gas flame the light scrap is heated very quickly to the temperatures close to its melting point. Then the scrap settles down and leaves the action zone of the flame which loses the convective contact with the scrap. In the course of the burners operation, the area of the heat-absorbing surface of the scrap lumps and the temperature difference between the scrap and the flame diminish progressively. The heat transfer remains high only during a short period after the start of the burners operation. Then the heat transfer reduces gradually and finally, drops so low that the burners must be turned off, as their operation becomes ineffective.

Besides, potential duration of conventional burners operation is also limited by the physical-chemical factors. At the scrap temperatures approaching 1400 °C and especially during the surface melting of scrap, the rate of oxidation of iron by the products of complete combustion of fuel sharply rises. In doing so, the products of fuel combustion are reduced to CO and H_2 according to the following reactions:

$$CO_2 + Fe = FeO + CO \qquad and \qquad H_2O + Fe = FeO + H_2$$

The fuel underburning increases, and CO and H_2 burn down in the gas evacuation system. The temperature of the off-gases rises sharply which, along with the other signs of reduced effectiveness of the burners operation, requires turning the burners off.

The described above processes in the scrap pile attacked by a narrow high-temperature flame explain comprehensively the futility of attempts to increase the unit power of conventional burners. Indeed, in accordance with well-known aerodynamic principles, the length and the volume of the flame and, therefore, its action zone increases insignificantly as the oxy-gas burner power increases. As a result, the critical temperatures causing fuel underburning and settlement of the scrap in this zone are reached in a shorter time. Respectively, approximately proportionally to the increase in power of the burner, the potential effective burner operation time is shortened, whereas the amount of heat transferred to the scrap increases insignificantly. Only a relatively small portion of scrap pile is heated, which has little effect on energy characteristics of the furnace.

Attempts to expand an action zone of the burners due to changing the shape of the flame during the heat are known. Both the narrow and flat or fan-shaped flames can be used in such burners depending on changeable conditions of the heating of scrap. The action zone of the fan-shaped flames as well as that of flat flames is expanded but their kinetic energy drops. An attempt to increase the power of such burners has been failed and they did not become spread.

Thus, in EAFs with conventional scrap charging from the top a possibility of reduction in electrical energy consumption due to the use of burners installed at the furnace freeboard is considerably limited and mainly exhausted. The continuous scrap charging allows carrying out the process of heating a scrap with the burners outside the furnace freeboard at separate devices which are much more fitted for this process. This significantly expands a possibility to save the electrical energy due to replacement of it with the energy of fuel.

1.6 Flat Bath Method with Continuous Scrap Charging

1.6.1 Key Features

As has been mentioned above, in case of conventional method of scrap charging from the top, the conditions of heating and melting of scrap by electric arcs, burners, and other energy sources vary drastically over the course of the heat, and the entire process has clearly defined non-stationary nature. Complete elimination of this deficiency which hinders further improvement of EAF's key characteristics cannot be achieved just by continuous scrap charging. It is essential that the continuously charged scrap is melted in the liquid metal only and that there is no scrap above the bath surface. If this condition is not observed, then scrap is melted both in the liquid metal and in the furnace freeboard. Then this method has all the major deficiencies of the conventional scrap charging from the top by one or several baskets. It should be noted that during this type of charging the substantial part of the scrap is also melted in the liquid metal.

Satisfying the above specified condition requires that scrap charging rate is equal to its melting rate in liquid metal. In this case the bath remains flat during the entire course of the heat. This process closely approaches to the continuous stationary process in which the conditions of scrap heating and melting do not change over the course of the heat. This allows maintaining metal temperature and electric power at a constant level.

1.6.2 Potentialities

The process of continuous scrap charging into flat bath has a number of essential potential advantages such as energetical, technological, and ecological. Let us outline these advantages and the factors which allow achieving these advantages.

• Increase of productivity and reduction of electrical energy consumption

The contributing factors are as follows: elimination of time spent on scrap charging from the top and of heat losses due to roof opening; improvement in conditions for stable maintenance of the foamed slag layer of the optimum thickness; increase in electrical energy efficiency due to complete submerging of arcs into the foamed slag during practically the entire heat; the usage of devices for the continuous scrap charging (conveyors, shafts) for scrap preheating by off-gases and by burners.

• Reduction of electrode consumption

Decrease of the freeboard height, especially in comparison with single charging furnaces, leads to reduction of electrode length and their side surface area. Oxidation of this surface by furnace gases is one of the major factors of electrode wearing. In the absence of tall scrap column in freeboard, the risk of electrode damage is drastically lower. All of this leads to the reduction of electrode consumption.

• Increase in yield

The factors affecting yield increase are an absence of scrap oxidation by furnace gases and by flames of the burners, as well as reduction of FeO in slag and of oxidation level of liquid metal due to the fact that ratio of concentrations of carbon and oxygen in the metal is close to the state of equilibrium. Furthermore, as off-gases pass through a conveyor tunnel or a shaft, the large and medium-size particles of dust containing iron precipitate and return to a bath with the scrap. Also, evaporation of metal does not occur, since the arcs do not contact solid scrap.

• Cost saving on electrical power supply and lowering of requirements for electrical grids

Due to elimination of the stage of unsteady burning of electric arcs during forming of cave in a column of scrap, there is reduction in voltage and frequency fluctuations generated by the arc furnaces in electric grids. The possibility of attaining a desired productivity using smaller transformer power allow for electrical grid of smaller power. All of this reduces costs on suppression of electrical interference in the grids as well as simplifies electric power supply to the furnaces.

• Increase of durability and service life of the wall and roof panels and other water-cooled elements

This is achieved by elimination of direct radiation of electric arcs which are being submerged in foamed slag at all times, which reduces significantly the risk of dangerous water leaks into the freeboard resulted from burnouts of the elements.

• Reduction of CO_2 emissions into the atmosphere and of expenditures on entrapping and extraction of dust from the off-gases: noise level reduction.

These ecological advantages are achieved due to elimination of non-controlled dust and gas emissions which take place in case of open roof and increase the quantity of cleaned gases by two or three times and also due to steady low-noise arcing in the foamed slag.

It should be emphasize that realization in practice of the above-mentioned potential advantages of continuous scrap charging into flat bath depends on key design features of the furnace using these concepts. Furnace design can cause some problems which hinder achieving the expected results. Overcoming these problems might require new original solutions. Related issues are discussed in the other chapters of the book with concrete examples of the furnaces using the elements of this new technology.

References

1. Kohl B, Sparlinek W, Ebner M (2012) Towards 2050—how the long term strategies of the European Union can affect EAF operations. In: 10th European electric steelmaking conference, Graz, Austria
2. Abel M, Hein M (2009) The SIMETAL ultimate at Colakoglu, Turkey. Iron Steel Technol 6(2):56–64
3. Sellan R, Fabbro M, Burin P (2008) The 300 t EAF meltshop at the new Iskenderun minimill complex. MPT Int 2:52–58
4. Morozov AN (1983) Modern steelmaking in arc furnaces. Metallurgiya, Moscow
5. Sanz A, Lavaroni G (2001) Advanced approaches to electric arc furnace offgas management. MPT Int 4:72–82
6. Gottardi R, Miani S, Partyka A (2008) A faster more efficient EAF. In: AISTech conference, Pittsburgh, USA

Chapter 2
Implementation of New Technology

Abstract Advantages and disadvantages of conveyor and shaft furnaces using the new technology and operating with flat bath are examined. Basic performances of these furnaces are compared with those of modern EAFs. Furnaces with flat bath having a number of important advantages, however, lag behind modern EAFs in productivity and in electrical energy consumption as well. In addition, shaft furnaces have very serious problems concerning environment.

Keywords Conveyor furnaces · Shaft furnaces with fingers · Shaft furnaces with pushers · Flat bath · Comparison of furnace performances · Scrap melting rate in liquid metal · Scrap charging rate · Environment regularities · Decomposition of dioxins and furans

2.1 Predecessors

Technology of continuous scrap charging into flat bath was developed and implemented by J.A. Vollomy who, in the late 1990s, built the first conveyor furnace Consteel [1]. In 2007, G. Fuchs adapted this technology for use in his shaft furnaces by placing the shaft with scrap alongside the EAF and by equipping the shaft with hydraulic pusher [2]. These breakthrough innovations as well as the other innovations have predecessors and very informative history.

In the late 1960s, M.A. Glinkov proposed and tried out continuous scrap charging in 600-t open-hearth furnaces at several plants in Russia and Ukraine. The furnaces operated on metal-charge with about 60 % hot metal content and with intense oxygen bath blowing. The new technology became known as scrap-oxygen process. The scrap was charged, as usual, by molds through the furnace windows after hot metal charging. Under the conditions of open-hearth furnaces, the new process did not demonstrate any advantages, but tremendously hindered production management of the open-hearth shops. That is why the try-outs of this method have been stopped and have never resumed.

© The Author(s) 2015
Y.N. Toulouevski and I.Y. Zinurov, *Electric Arc Furnace with Flat Bath*,
SpringerBriefs in Applied Sciences and Technology,
DOI 10.1007/978-3-319-15886-0_2

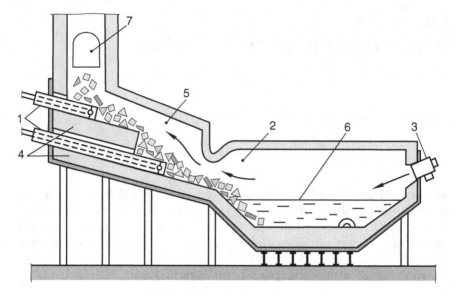

Fig. 2.1 The Tring's furnace

In 1961 in West Germany, M. Tring created direct flow recuperative furnace with productivity 10 t/h for semi-finished product melting [3]. The furnace design included the elements resembling some fundamental key features of modern conveyor and shaft furnaces with continuous scrap charging. The scrap was moved by pushers (1) to the melting zone (2), equipped with burners (3) on the sloping stepped bottom (4) of the chamber (5) adjacent to the bath (6), Fig. 2.1. In the process, the scrap was heated by gases being removed from the melting zone (2) through the chamber (5) into the gas duct (7). Air for the burners was preheated in the recuperator. Industrial tests have revealed essential shortcomings of these furnaces and they have not been developed further.

In 1980s–1990s, so called energy optimizing furnaces (EOF) developed by Pains have been operating in Brazil [4]. The freeboard shape of these furnaces resembled that of EAF except for the fact that they did not have electrodes since they used energy of fuel instead of electrical energy. The shaft scrap preheater was placed above the central part of water-cooled furnace roof. The off-gases were removed from the freeboard through this shaft, Fig. 2.2. The shaft was divided from top to bottom into several chambers by the movable water-cooled fingers. A portion of scrap was placed on the fingers in each chamber. The total mass of scrap in all chambers was equivalent to the amount of scrap needed for one heat.

The burners were installed under the lower chamber. These burners provided the additional amount of heat required for high-temperature scrap heating. When the fingers in any of the chambers moved apart, the scrap dropped onto the fingers of the chamber below. The scrap from the lower chamber dropped into the furnace.

Fig. 2.2 Sectional shaft
scrap preheater on EOF
furnace *1* shaft; *2* finger;
3 oxy-fuel burners

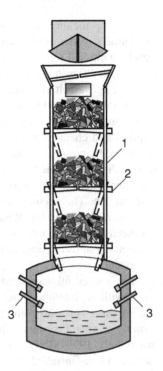

The partitioned shaft scrap preheater makes the EOFs quite close analogs of modern conveyor and shaft furnaces. The similarity of EOFs and conveyor EAFs is even stronger due to the fact that the shaft preheater divided into several chambers is, essentially, a kind of the vertical conveyor which enables charging of scrap into the furnace by separate small portions following each other.

The preheater functions in the following manner. When each new heat starts, all the scrap heated during the previous heat is already in the preheater. The scrap in the lower chamber is preheated to the maximum required temperature, whereas in the upper chambers the temperature is decreasing. The scrap in the top chamber has the lowest temperature. The heat starts with discharging scrap from the lower chamber into the furnace. Then the scrap is transferred, in order, from each chamber into the next chamber below. The scrap transferred into the lowest chamber is being further heated there during a certain short period of time, while the empty uppermost chamber is being charged with the cold scrap. As soon as the scrap in the lower chamber is preheated to the maximum required temperature, it is discharged into the furnace, and the next cycle of the scrap transfer from the top to the bottom and charging of upper chamber with a new portion of scrap is repeated. The total required for a heat amount of scrap preheated to the maximum required temperature is charged into the furnace in several cycles, depending on the number of chambers. Concurrently with finishing of charging the scrap into the furnace, the preheater is filled with scrap for the next heat.

For scrap melting and, to a far lesser degree, for liquid metal heating, oxygen-fuel burners were used in EOFs. Oxygen was used not only in the burners, but for bath blowing and for CO post-combustion in freeboard as well. Total oxygen flow rate was 60–80 m³/t. Fuel consumption (expressed in terms of fuel equivalent) was about 10 kg/t. Such low fuel consumption could be explained by scrap preheating to 800–850 °C as well as by high content of hot metal in charge.

Heating of metal to tapping temperatures was carried out mainly by means of physical and chemical heat of hot metal; its content in metal charge was usually 50–60 %. Increasing a share of scrap in a charge to more than 50 % resulted in serious difficulties due to very low effectiveness of heating of liquid bath by burners. There was sharp increase in tap-to-tap time which on EOF usually amounted to about 1 h. The attempts to process in the EOFs a charge consisting mainly of scrap were unsuccessful. Relatively low productivity of the EOFs, a need to use hot metal in large quantities, and some other factors made it impossible for these steelmaking units to compete with EAFs. As a result, they were used to a very limited extent.

More successful was an attempt to implement continuous scrap charging combined with high-temperature scrap heating by off-gases in a BBC-Brusa steelmaking unit. This unit combines an EAF with a rotary tube-type heating furnace forming a single system. The first installation of this kind started operating in the early 1970s (Italy) [5]. In this unit, a 13 m long rotary tube-type heating furnace (1) was installed above a 36-t EAF, Fig. 2.3.

The gases escaping through an opening in the EAF roof are drawn into the tube-type furnace. The scrap is continuously charged into the bath of the furnace through the same opening. When passing through the tube furnace, the gases are heating the scrap coming from a batcher (2) equipped with vibrator (3). At the upper end of the furnace the cooled gases are drawn into a fume hood (4) and are removed for purification.

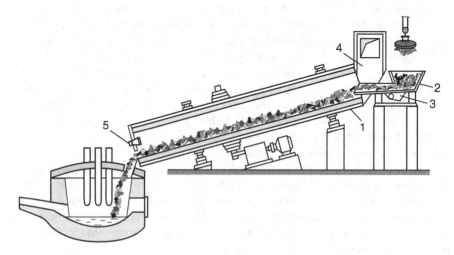

Fig. 2.3 BBC-Brusa unit (designations are given in the text)

Gas burners (5) are located at the lower end of the rotary furnace. The scrap passes through the furnace for 6–10 min. During this time the scrap is heated up to medium mass temperature of about 1000 °C. This temperature was reached not quite due to off-gas heat but mostly due to the burners which account approximately 73 % of all heat coming into the rotary furnace. Rotation of the furnace prevents welding of the scrap lumps despite their high temperature and assures that heat from the refractory lining is being used for scrap heating. This enhances the advantages of countercurrent system of gas and scrap motion which makes gases exit the furnace at low temperature. The thermal efficiency of the rotary furnace calculated for the total heat input reached approximately 45 %.

Performance of the BBC-Brusa unit equipped with transformer power of 7.2 MW for 3 years of service shows great potentialities and principle energy advantages of high-temperature scrap heating. With natural gas consumption of 30 m^3/t, electric energy consumption was cut by 220 kWh/t. Furnace productivity increased up to 100,000 ton per year which at that time was equal to the productivity of a furnace with the same capacity, but equipped with high-power transformer. Continuous charge of scrap assured very quiet arcing and low noise level (less that 80 db). Durability of the refractory lining of the rotary furnace was 1500 heats.

Despite the advantages attributed to the high-temperature scrap heating, such units had quite limited use and only for a short period of time. This can be explained by the fact that for the modern high-productivity EAF the dimensions of a rotary furnace required call for really too big size and height of the buildings for EAF's shops. Besides, rotary furnaces can operate only using properly prepared fragmentized scrap. This narrows raw material supply base and increases cost. The units with rotary furnaces also have other significant drawbacks which prevent them from being used. Nevertheless, the impressive results obtained on BBC-Brusa units promoted a search for new options of high-temperature heating a scrap in combination with continuous charging it into the bath. This has resulted in development of modern shaft and conveyor EAFs.

2.2 Conveyor Furnaces

2.2.1 Design and Technological Process

Due to persistent efforts of Tenova company, conveyor electric arc furnaces Consteel designed by J.A. Vollomy have become considerably widespread. On the late 2013, there are over 40 of these furnaces worldwide. This steelmaking unit combines an electric arc furnace and vibratory conveyor (2) consisting of chutes made of sheet steel. A conveyor is adjacent to a side wall window of an arc furnace from the side opposite to a furnace transformer, Fig. 2.4.

Total length of the conveyor is about 100 m. A part of the conveyor is inside of a refractory-lined tunnel (1) adjoining the furnace. The length of this tunnel is about 30 m. The off-gases leaving the furnace are removed through this tunnel. In the

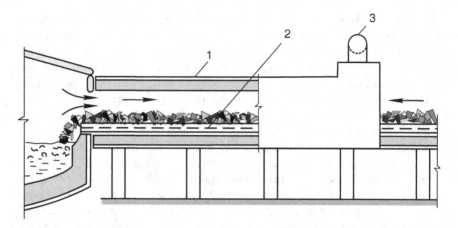

Fig. 2.4 Consteel system (designations are given in the text)

tunnel, the gases move in the opposite direction to the scrap and heat it up. Then the gases are directed through the water-cooled duct (3) into the bag filters for dust extraction. The conveyor chutes located in the tunnel are cooled by water.

Scrap charging into a bath is carried out by special water-cooled chute installed at the end of the conveyor. This chute is filled with scrap at regular intervals, then moved into a freeboard through a window, and the scrap is dropped into a bath. The width of the conveyor depends on furnace capacity and on the 350-t furnaces is about 2.5 m. A furnace roof is opened only during the first heat after a furnace scheduled maintenance before which all liquid metal and slag is tapped from the furnace. During this heat, one basket of scrap is charged into the furnace from the top in order to accumulate on the bottom a sufficient amount of liquid metal to start continuous scrap charging by conveyor. During all other heats, the furnace operates with a hot heel and charging is carried out by conveyor only without opening the roof. This significantly reduces dust and gas emissions into the atmosphere of the shop.

The Consteel furnaces do not impose much stricter requirements to the quality of scrap preparation for the heat in comparison to the conventional EAFs with scrap charging from the top. The rate of scrap charging into the bath is always kept equal to the rate of scrap melting in liquid metal. As a result, the bath remains flat during the whole melting process. During this period, the temperature of metal is kept at a constant level of 1560–1580 °C. Temperature increase above this level is not recommended in order to avoid sharp drop in durability of the refractory lining of the bottom and the banks of the furnace.

The rate of scrap melting in liquid metal is determined by intensity of convective heat transfer from metal to scrap which depends first and foremost on speed of metal streams flowing over the surface of scrap pieces as well as on difference in temperatures between metal and scrap. Amount of heat Q, kW/t, transferred from liquid metal to scrap per unit time is defined by the following equation:

$$Q = \alpha \times F \times \Delta t \tag{2.1}$$

α, kW/(m^2 °C)—coefficient of convective heat transfer from liquid metal to scrap.

Δt, °C—average difference in temperatures between metal and scrap pieces, per melting period.

F, m^2/t—specific surface area of all the scrap pieces submerged into liquid metal.

In Consteel furnaces, electric arcs are constantly submerged into foamed slag. They do not have direct heat contact with scrap and do not affect intensity of heat transfer from metal to scrap. The arcs participate in scrap melting only indirectly by maintaining metal temperature at a required constant level. Given a specific furnace capacity, such melting mechanism precludes the possibility of increasing melting rate by increasing the power of arcs. The latter must strictly correspond to the scrap melting rate which does not depend on the power of arcs. Thus, the power of arcs does not determine scrap melting rate, but, on the contrary, scrap melting rate determines the maximum permissible power of arcs which must not result in overheating of metal. In the EAFs, such a limitation of the electric power does not exist. As the arc power rises, both melting rate and hourly productivity of the furnaces increase.

In the modern EAFs, the scrap melting process may be divided into two periods. During the first period, the main body of scrap pile is melted down with the electric arcs in the furnace freeboard above the hot heel surface. Plasma of the arcs has a temperature of more than 5000 °C and possesses high kinetic energy. The heat energy of the arcs is transferred to the scrap by both radiation and convection. The intensity of these heat transfer processes is much higher than that in the liquid metal where the difference between a temperature of the liquid metal and scrap melting point amounts to 30–40 °C only. The power of the arcs is also higher than that of the Consteel furnaces. Furthermore, the oxy-gas burners and the process gases formed during post-combustion of CO in the freeboard contribute to the heating and melting of scrap. At the first period of the scrap melting process in EAFs, all these factors must ensure the higher melting rate in comparison with that in Consteel furnaces.

During the second period, upon completion of both forming of the flat bath and submerging of the arcs into a foamed slag, the rest of the scrap melts down in the liquid metal. If any, the mechanism of heat transfer to scrap is the same as in the Consteel process. However, just like in the first period, the rate of melting has to be considerably higher. It could be explained by the fact that the scrap lumps before submerging into the melt are heated up to a temperature close to the melting point in the EAF freeboard. Only a small portion of fine scrap placed on the bottom of the basket submerges into the hot heel not being preheated. Increase of melting rate must result in the higher productivity of the moderns EAFs in comparison with the conveyor furnaces. These considerations are confirmed by actual data.

2.2.2 Comparison Between Conveyor Furnaces and Modern EAFs

The key performances of the furnaces were compared: hourly productivity and electrical energy consumption, as well as scrap melting rate which to a large extent determines productivity. The furnaces using hot metal were not considered since their performances are determined by hot metal percentage in the charge rather than by the design features and operating modes of the furnaces, Chap. 1, Sect. 1.2.

Despite the fact that there are more than 40 conveyor furnaces operating world-wide and that numerous articles related to these furnaces have been published by Tenova personnel in such magazines as "MPT International" and "Iron & Steel Technology", in Proceedings of the European and US conferences as well as in other information sources, data allowing to determine scrap melting rate and hourly productivity are given only for several conveyor furnaces operating without hot metal. Even less data are provided on electric energy consumption. This allows to suppose that published data show the very best achievements of the conveyor furnaces. Therefore, comparison between these furnaces and the state-of-the-art modern EAFs seems to be most objective.

This comparison appears to be the most indicative with respect to estimating the potential of the steel melting units being analysed. All the data corresponding to the above-indicated requirements have been used for the purpose of this comparison. Advertising data not containing the actual performances of the furnaces operation have not been taken into consideration.

The rate of melting was determined by dividing the total mass of scrap and pig iron by melting time. In the cases when the data on composition of the charge were absent, the mass of solid metal-charge was determined using the tapping weight with consideration of the liquid steel yield. The melting time was determined using the power-on time with due correction for the time needed for heating of the melt to the tapping temperature. Despite very poor data used for making a comparison of furnaces the curves plotted based on these data are characterized by an insignificant spread of points which allows to draw the quite definite conclusions.

With the increase in the capacity M^1 the scrap melting rate S increases in both furnace types, Fig. 2.5. This could be explained by an adequate increase in electrical power of the furnaces. In the Consteel furnaces, with the increase in their capacity the weight of hot heel grows which allows speeding up the scrap charging rate and, consequently, increasing the power of arcs.[2] For the entire range of the furnace capacities from 100 up to 350 tons, the rate of scrap melting in the EAFs

---

[1] The furnace capacity M was determined by the sum of weights of both the hot heel and tapping.

[2] An effect of the hot heel weight as well as other factors on a permissible scrap charging rate, equal to the melting rate, is reviewed in detail in Chap. 3.

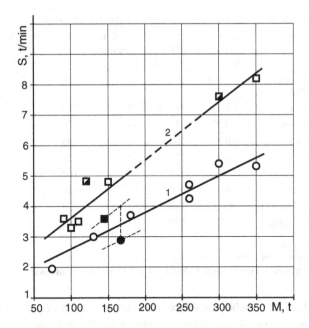

Fig. 2.5 Scrap melting rate S, t/min versus furnace capacity M, t *1* Consteel furnaces [1, 6–10]; *2* EAFs [11–17]. ○ Consteel furnaces including ● those in Russia, ☐ EAFs including ◪ design data, ■ in Russia

is higher than that in the conveyor furnaces by approximately 1.6 times. In addition, the ratio of the melting rate in the modern EAF to that in the conveyor furnace operating under similar conditions at the same region in Russia after reducing them to the equal M amounts to 1.6 as well, Fig. 2.5.

The Consteel process eliminates times expended on upper scrap charging with baskets which shortens power-off time. Due to this advantage, when replacing EAFs with Consteel furnaces, the total shortening tap-to-tap time and increasing hourly productivity were reached in a number of cases despite reduction in the melting rate. At present, the situation changed. Owing to an increase in the capacity of the new generation furnaces, more and more frequently scrap is charged by a single basket in EAFs of low and medium capacities and by two baskets in EAFs of high capacity. This fact along with speeding-up of crane mechanisms and furnace drives, training of furnace personnel and improving in work coordination has allowed to shorten power-off time in the best EAFs down to a level of that in Consteel furnaces, Chap. 1, Sects. 1.3.7 and 1.5.1. The combination of these achievements with greater scrap melting rate ensures increase in productivity of the EAFs by about 20–25 % on average, as compared to the conveyor Consteel furnaces, given equal capacities, Table 2.1.

Electrical energy consumption in the conveyor furnaces is approximately the same as or even higher than that of the EAFs operating without scrap preheating. According to the data in Table 2.1, it is about 8 % higher. Electrical energy consumption is

Table 2.1 Performances comparison between EAFs and consteel furnaces

Furnace type	Country	Capacity/ tapping weight, t	Power MVA	Tap-to-tap time min	Productivity t/h	Electrical energy kWh/t
EAF	Europe: USA	120–130	150–160	<35	>200	345–355
	Russia	146/128	95	55	140	374
	Turkey [17]	380/320	240	60	320	359
	USA [11]	110/82	110	32	155	317
Consteel	Norway [7]	123/83	75	41	122	384
	Greece [8]	180/130	120	48	162	395
	Russia	168/118	90	61	116	416
	Thailand [1]	300/187	125[b]	65	173	363[c]
	Italy [10]	350/250[a]	209	63	238	–

[a]Tapping weight, expected
[b]The values calculated per active powers of 95, 34 MW
[c]For 40 % of pig iron in a charge which reduces electrical energy consumption

closely related to scrap preheating temperature. When Consteel furnaces were being developed, it was expected that scrap on conveyor would be preheated by off-gases up to 700–900 °C. Such preheating could have ensured quite considerable electrical energy savings. However, these expectations were not realized. Scrap preheating in Consteel furnaces proved to be ineffective. Mass average temperature of scrap preheating does not exceed 250 °C on the medium capacity furnaces and is even lower on the 350-t furnaces because of the increased scrap layer thickness (reaching 900 mm). According to energy balance of one of these furnaces, the enthalpy (in other words, heat content) of the preheated scrap E_s is 16.4 kWh per ton of scrap [18]. Interpolating between the values of temperature t_s equal to 100 and 150 °C, Table 1.1, Chap. 1, we find that the enthalpy $E_s = $ of 16.4 kWh/t corresponds to the temperature of scrap $t_s = 125$ °C. Low effectiveness of scrap preheating in the Consteel furnaces is explained by an unsatisfactory regime of heat transfer from gases to scrap in the heated tunnel as well as by insufficient heat power of off-gases flow. These factors are closely examined in Chaps. 3 and 4.

In countries like Russia, Norway, etc., operating the Consteel furnaces during winter time causes considerable difficulties resulting from low scrap preheating temperatures. Getting into conveyor, snow and ice melt in the tunnel heated by gases. Water formed as a result does not have time to evaporate. It flows into the lower part of the conveyor chutes, mixes up with mineral debris contained in the scrap, and forms mud deposits which, along with the water and wet scrap, are charged into the bath of the furnace. This results in dangerous "popping" and intense metal splashing, and, in case of unfavourable combination of circumstances, can even lead to explosions with catastrophic consequences, which is confirmed by the experience of operating the 170-t Consteel furnace, Asha, Russia. Scrap preheating temperature on this furnace is 150–250 °C [19]. Due to such low scrap preheating temperatures an amount of dioxins emitting in this case is insignificant and, therefore, the problem associated with decomposition of dioxins in operating Consteel furnaces did not arise so far.

Along with afore noted shortcomings Consteel furnaces in comparison with EAFs have important advantages thanks to flat bath operating. These advantages are common for all the furnaces operating with flat bath including shaft furnaces as well. Therefore, the latter is examined at the end of the chapter, Sect. 2.4.

2.3 Shaft Furnaces

2.3.1 Furnaces with Fingers Retaining Scrap

The development of shaft furnaces is associated with the name of G. Fuchs. In the first 90-t shaft furnace the water-cooled shaft was installed above the furnace roof and did not have fingers retaining the scrap in the shaft. The scrap was charged into the furnace through the shaft. While the lower part of the scrap pile was located on the bottom of the furnace, its upper part was in the shaft. Gases from the furnace were evacuated through the shaft and heated the scrap located in it. As the scrap melted in the furnace, the entire scrap pile settled down. This created the free space in the shaft, which allowed charging of additional portions of the scrap.

Later G. Fuchs has developed and put into operation at several plants the furnaces with one row of fingers in the lower part of a shaft, Fig. 2.6. The scrap is charged into the furnace by two baskets. At the tapping the scrap of the first basket heated by the off gases during the previous heat lies on the fingers in the shaft. After the tapping the fingers split apart, and the heated scrap is charged into the furnace. After that, the cold scrap from the second basket is charged into the empty shaft. The share of the scrap from the second basket remaining in the shaft is heated by the off gases passing through the shaft. As the scrap melts in the furnace, the scrap in the shaft rapidly caves in and the shaft clears. The fingers are then shut, and the first basket of scrap for the following heat is charged on the fingers. By the tapping time this portion of scrap is already preheated by the off-gases up to relatively high temperature.

With this heating method when gases pass through the scrap from the bottom to the top, the overheating and even the partial melting of the lower scrap layer do not create any problems. The melt and the liquid slag formed flow down into the furnace and do not obstruct splitting fingers apart and scrap discharging. At those periods of the heat when the off-gas temperature reduces, a lack of heat can be compensated by using of burners installed under the shaft. Heating a scrap layer on the fingers with gases passing through the layer is much more efficient than surface heating a scrap on a Consteel conveyor. All this contributes to increasing the average mass temperature of scrap heating and reducing electrical energy consumption. Another advantage of shaft furnaces is that a substantial part of the dust carried out from the freeboard settles down in the layer of scrap. Due to this fact the yield is increased by approximately 1 %.

Fig. 2.6 Shaft furnace with
fingers retaining scrap (before
the beginning of the heat)

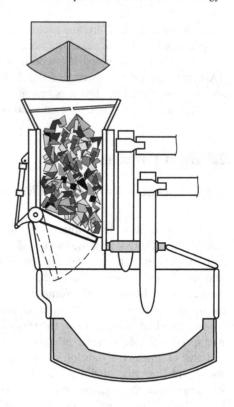

Although there is no reliable data on average-mass temperatures of scrap heating for the finger shaft furnaces, this temperature can be estimated with the help of the heat balance data for one of such furnaces. In accordance with these data for the 135-t shaft EAF with the 120 MVA transformer and tap-to-tap time of approximately 38 min, the first basket of heated scrap introduces 55 kWh/t of steel [20]. The total amount of scrap per heat is 148 tons. Assuming that the first basket contains 89 t (60 %) of scrap, we can find its enthalpy E = 55 × 135/89 = 83.4 kWh/t of scrap. Corresponding to this enthalpy is the average-mass temperature of scrap of 550 °C, Chap. 1, Table 1.1. It can be assumed that only half of the scrap from the second basket, i.e. 30.0 t, is in the shaft and that this additional amount of scrap is heated by the off-gases to the temperature of 550 °C as well. In this case, the average enthalpy of the entire amount of scrap heated by the off-gases will be (89.0 + 30.0) × 83.4/ 148 = 67.0 kWh/t of scrap, which corresponds to preheating temperature of 450 °C, Chap. 1, Table 1.1. Such a relatively low temperature can be explained by the facts that only a part of the scrap equal to approximately 80 % of its total amount is heated by the off-gases, the duration of heating of the scrap from the first basket on the fingers is short, and also heat power of the off-gas flow is relatively low.

In this case, heating by the off-gases returns to the process 83.4(89 + 30)/ 135 = 73.5 kWh/t of steel. Assuming the electrical energy efficiency coefficient η_{EL} = 0.75, we will find that reduction of the required useful heat consumption by 73.5 kWh/t can reduce electrical energy consumption by $73.5/0.75 = 98$ kWh/t. The minimum electrical energy consumption on the shaft furnace under consideration [20] was 285 kWh/t. In comparison with conventional EAFs operating without scrap preheating, the reduction of electrical energy consumption is about 90 kWh/t.

Later design of a finger shaft furnace was persistently improved. The latest variation of the furnace, named Siemetal EAF Quantum, developed by Siemens VAI Metals Technologies, Germany, along with other innovations comprises a new system of shaft charging and improved design of fingers retaining scrap [21].

Scrap is charged into the shaft from above with a special chute moving up and down on inclined elevator rather than baskets with a crane. The chute is loaded on scrap yard and contains one third of the scrap required for the heat. Such a system allows covering the shaft with a hood and considerable decreasing uncontrolled gas-dust emissions from the shaft during the charging of scrap.

As opposite to the fingers shown in Fig. 2.6, in the EAF Quantum the fingers, similar to pitchforks, are introduced into the shaft through its sidewalls. To charge a batch of the preheated scrap into the bath the fingers are pulled out of the sidewalls of the shaft. After charging they are immediately introduced into the shaft to receipt the next batch of scrap. Thus, all the scrap is heated on the fingers what allows increasing its average mass temperature. The prior system, Fig. 2.6, did not enable to close the fingers at once after falling of the first batch of scrap into the furnace freeboard. It was necessary to wait for caving-in of scrap. Therefore, only the first basket of scrap was heated on the fingers.

Designed productivity of a 100-t Quantum furnace with a transformer of 80 MVA is 182 t/h when charging 3 chutes, and when charging 4 chutes is 162 t/h. Expected electrical energy consumption is 280 kWh/t [21]. Since approximately even electrical energy consumption has been achieved at the Fuchs' finger shaft furnace in the late 1990s [20] it can be supposed that in the new furnace an average mass temperature of scrap preheating with off-gases will not exceed 450–500 °C.

2.3.2 Shaft Furnaces with Pushers

The off-gas heat efficiency when heating a scrap in a shaft is higher by approximately three times than that when scrap is heated on a Consteel conveyor. On the other hand, continuous furnace operation with the flat bath is an advantage of the Consteel process. To combine these both advantages G. Fuchs has developed and implemented real-life shaft furnaces with the continuous charging of scrap into the liquid bath. The design of such a furnace named COSS is schematically shown in Fig. 2.7.

A rectangular shaft (1) is installed on the cart (2) next to the furnace. The shaft is connected to the furnace with a short tunnel (3). A sliding gate (4) opens for charging of scrap into the shaft. The charging is carried out with the power-on and does not interrupt the furnace operation. A gas duct (5) is placed under the sliding

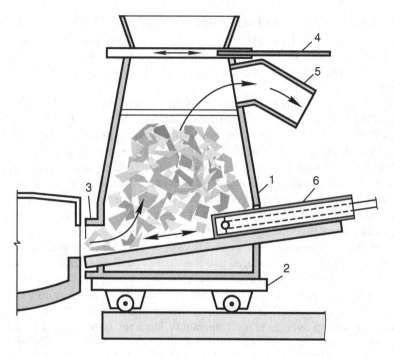

Fig. 2.7 Shaft furnace with a pusher. Continuous scrap charging into the bath (designations in the text)

gate (4). The shaft is lined with the massive steel segments and has no water-cooled elements which could be damaged during the scrap charging. The mass of scrap in the shaft is gauged by the measuring elements on which the shaft rests. This makes it possible to control the rate of charging of scrap into the furnace.

The scrap is charged continuously into the liquid bath with the help of the pusher (6) which is moved forth and back by the well protected hydraulic cylinder. During the entire period of charging and melting of scrap in the liquid metal its temperature is kept by the electric arcs at the constant level of 1560–1580 °C, just like in the Consteel process. All the fundamental features of this method of scrap melting, its advantages and shortcomings are the same as those of the Consteel furnaces. Unlike in the first finger shaft furnaces, in the furnaces with the pushers the off-gases heat all the scrap. This made it possible to expect the higher temperatures of heating. Unfortunately, the reliable data on the temperatures for the furnaces operating without hot metal are absent.

As reported by G. Fuchs,[3] the hourly productivity of the three 140–150-t furnaces put into operation recently was 125–149 t/h with the minimum electrical energy consumption of 300–341 kWh/t on the better days. With regard to productivity, these furnaces are considerably inferior to the modern EAFs

---

[3] Presentation: Ekaterinburg, Russia, 2012.

which can be explained not only by the relatively low melting rate of scrap in liquid metal, but by the low power of transformers (80 MVA) as well. The electrical energy consumption does not differ significantly from that of the best results achieved on the modern EAFs operating without scrap preheating during some of the short periods of their operation. Obviously, that the high-temperature scrap preheating has not been achieved in these shaft furnaces either. This can be associated with both the insufficient duration of the heating and relatively low heat power of the off-gases flow in furnaces operating without use of hot metal.

When operating with hot metal, the heat power of the off-gas flow increases sharply what makes it possible to achieve the high-temperature heating of scrap in the shaft furnaces. As an example, let us review the performances of the shaft furnace of 150-t capacity with pusher operating in China [2]. For this furnace, share of hot metal in the charge is 40 %, tap-to-tap time is 35 min, oxygen flow rate is 40 m^3/t, and electrical energy consumption is less than 100 kWh/t.[4] Scrap melting is carried out with low electrical energy consumption, mainly by means of sensible and chemical heat introduced with hot metal, by exothermic reactions of oxidation of C, Si, Mn, and P, and by highly heated scrap which temperature reaches 800–1000 °C. Operating experience proves that the pusher works reliably at such temperatures of scrap. It is worth mentioning that such performances can also be achieved on the EAF without scrap heating if the share of hot metal in a charge is significantly increased.

Elimination of gas-dust emissions when charging scrap into the shaft is one of topical tasks of improving of shaft furnaces operation with continuous scrap melting in the liquid bath. In order to solve this problem an original scrap charging system named EPC [22] has been developed by the Company KR Tec GmbH, Germany. Operating principle of this system is explained by Fig. 2.8.

The main component of the system is a movable hopper (1) with an opening bottom placed in a bunker (2) adjacent to the shaft (3). By means of hydraulic cylinders of the bunker the hopper can be moved into the shaft through an aperture in its sidewall. When the hopper is placed in the bunker its front wall is closing the aperture. When the hopper is positioned in the shaft the aperture is closed with the hopper back wall. At this position the bottom of the hoper is opened and the scrap falls smoothly into the shaft.

In the course of feeding of the scrap into the bath by means of a pusher (4) the scrap caves in inside the shaft and the hopper can be moved backward to the bunker. The hopper is charged there again with the help of a scrap basket (5). A slide gate (6) is opened for a while to charge the hopper only. Thus, during the heat the scrap can be charged into the shaft by separate batches without loss of airtight of the system. Therefore, gas-dust emissions from the shaft into the shop atmosphere are almost completely eliminated.

---

[4] This figure seems somewhat understated.

Fig. 2.8 EPC system of scrap charging into the shaft (designations are given in the text)

Melting of the scrap starts when the scrap batch preheated in the shaft during the prior heat is feeding into the hot heel. The system allows realizing different variations of furnaces operation. During the heat one or several of scrap batches can be charged into the shaft depend on a furnace capacity, shaft and hopper dimensions. This requires the certain number of hopper movements and charging of the hopper with baskets. Figure 2.8 shows a stage of the heat preceding the next charging of the scrap from the hopper into the shaft. Off-gases passing through the scrap layer heat it and are evacuated via a gas duct (7).

2.3.3 Ecological Problems

During development and implementation of the shaft furnaces, the following problem occurred: personnel and environment had to be protected from highly toxic compounds of halogens and hydrocarbons (usually referred to by the generic term "dioxins") contained in the off-gases. When the regular grades of steel are produced in EAF, relatively cheap scrap contaminated by plastic, rubber, upholstering materials from car interiors, and oil is used. When such scrap is preheated in shafts by off-gases to temperatures exceeding 400 °C these contaminators burn down with the formation of dioxins. The gases leaving the shaft with temperatures of the order of 400–500 °C are saturated with dioxins. For complete decomposition of dioxins, further heating of these gases by burners in special chambers to approximately 1000 °C and even higher is needed. This involves additional natural gas flow rate of 5.5 m^3 or more per ton of steel, which sharply reduces the energy efficiency of the process.

As already mentioned, the Consteel furnaces do not encounter the problem of dioxins only because the scrap on the conveyor is preheated to quite low temperatures at which the formation of dioxins is insignificant. It is known that dioxins are practically absent in case of smelting of alloy and special grades of steel for which clean scrap is used. However, due to economic considerations, for mass production of EAF steel, it is impossible to avoid the use of cheap contaminated scrap as well as to ensure its thorough cleaning.

In North America, Western Europe, and some other countries, the acceptable concentration of dioxins in the atmospheric emissions is limited by legislative regulations and cannot exceed extremely low values of the order of 10^{-10} g/m^3. Since removal of dioxins from gases involves significant additional power consumption, the shaft furnaces have spread to a limited extent only, mainly in China, Indonesia and other countries, where such strict regulations so far do not exist. However, such an approach to solving this problem is unpromising.

In EAF, dioxins are also formed after charging of each next basket of scrap. However, under the conditions of high temperatures of the freeboard and of the entry part of the gas duct, dioxins decompose completely. Then, the gases cool down as they move through the duct, and at the temperatures below 600 °C decomposed dioxins may reform. This process takes some time and occurs only if the gases cool down relatively slowly. To avoid reforming of dioxins, atomized water is injected into the gas flow downstream. Water evaporates quickly, and the temperature of the gases sharply drops to approximately 200 °C, at which reforming of dioxins is completely avoided. Thus, suppression of dioxins formation in EAF does not require additional power consumption, as opposed to the shaft furnaces.

It is important to emphasize that water injection into the gas duct not only prevent reforming of dioxins, but also completely eliminates possibility of further post-combustion of carbon monoxide CO. Allowable CO emissions into atmosphere are also strictly limited in many countries. Therefore, when water injection is used, it is essential to achieve complete post-combustion of CO within the duct before the point of injection; this ensures that allowable CO emissions into atmosphere are not exceeded. In modern gas evacuation systems this problem is solved by intensifying the process of mixing of furnace gases flow with flow of the air drawn into duct.

2.4 Results of the Implementation

The principal potential advantage of the technology of flat bath with continuous scrap melting in liquid metal is a possibility to significantly increase productivity and to reduce electric energy consumption, Chap. 1, Sect. 1.6.2. In the modern conveyor and shaft furnaces this advantage has not been realized so far. With regard to both productivity and electric energy consumption, these furnaces significantly trail the modern EAFs using the conventional methods of scrap charging and melting. If this challenge is not overcome, the conveyor and shaft furnaces will not be able to successfully compete with the EAFs, replace them, or to be considered as steel melting units of the future.

In case of equal capacities, the power of transformers of the conveyor and shaft furnaces is considerably lower than that of the EAFs, which allows to install these furnaces in the regions with electrical supply grids of relatively low power. However, this could be considered as an advantage of the conveyor and shaft furnaces only in case of identical productivity of these furnaces with that of the EAFs. Otherwise, the lower electrical power resulting from relatively low melting rate of scrap cannot be considered as advantage because installation of the EAF of the same power ensures higher productivity. However, the furnaces with flat bath actually impose lowered requirements to electrical grids. This advantage remains in case of transformer power equal to that of the EAF. Such an advantage can be explained by the fact that the furnaces with flat bath have a lowered level of electrical interferences generated in the grids due to improving arcing stability.

Almost all the other potential advantages mentioned in Sect. 1.6.2, Chap. 1, have been also realized fully or partially. Furthermore, due to relatively low temperature of metal bath as well as to possibility of slag replenishment during scrap melting stage in the furnaces with flat bath, better conditions for dephosphorization are created as compared to the conventional EAFs. At the same time, increasing the content of FeO in slag is not required, which contributes to an increase in yield. All these advantages ensured quite wide spread of the conveyor furnaces. The ecological problems caused by forming of dioxins hindered spreading of the shaft furnaces, Sect. 2.3.3.

In order to determine the most efficient methods of increasing productivity and reducing electrical energy consumption on the conveyor and shaft furnaces, the detailed analysis of the processes of scrap melting in liquid metal is needed. Chapter 3 addresses this analysis.

References

1. Memoli F, Guzzon M, Giavani C (2011) The evolution of preheating and importance of hot heel in supersized Consteel® systems. In: AISTech conference, Indianapolis, USA
2. Fuchs G, Rummler K, Hissig M (2008) New energy saving electric arc furnace designs. In: AISTech conference, Pittsburgh, USA
3. Tring M (1966) The furnace for continuous scrap melting. J Iron Steel Inst 44–50
4. Bonestell JE, Weber R (1985) EOF (energy optimizing furnace) steelmaking, Iron Steel Eng 62(10):16–22
5. Neumann F, Leu H, Brusa U (1975) BBC-Brusa is steelmaking process. Stahl Eisen 95(1):16–23
6. Miranda U, Lombardi E, Bosi P (1999) Saving energy and protecting the environment the first Consteel plant in Europe. In: Proceedings, 6th European electric steelmaking conference, Düsseldorf, p 16
7. Giavani C, Guzzon M, Picciolo F (2010) Start-up and results of the EAF Consteel plant of Celsa ordic-Mo i Rana (Norway). In: Proceedings, AISTech conference, vol 1. Pittsburgh, p 783
8. Bouganosopoulos G, Papantoniou V, Sismanis P (2008) Start-up experience and results of Consteel at the SOVEL meltshop. In: Proceedings, AISTech conference, vol 1. Pittsburgh, p 124
9. Andrews S, Powers J, Fox M et al (2003) Start-up and operating results of the Cojet® sidewall injectors on a Consteel furnace. In: Proceedings, AISE annual convection, Pittsburgh, p 49
10. Ferri MB, Giavani C (2009) World's largest Consteel® plant feeds the new thin slab mill at Arvedi, Italy. MPT Int (6):26–28
11. Cottardi R, Miani S, Partyka A et al (2008) Faster, more efficient EAF. In: Proceedings AISTech conference, vol 1. Pittsburgh p 205
12. Patrizio D, Pesamosca A (2012) Successful Start-up and benchmark operating results of 100-t FastArc EAF at Kosco, Korea. In: Proceedings, AISTech conference, Atlanta, p 777
13. Salomone Ph, Legrand M, Fabbro M (2004) Increasing productivity at the Duferco La Louviere meltshop. MPT Int (2):52–56
14. Alzetta F, Poloni A, Ruscio E (2006) Revolutionary new high-tech electric arc furnace. MPT Int (5):48–55
15. Fabbro M, Chiogna A, Cappellari G (2009) Steelmaking and casting at the new minimill of Siderurgica Balboa. MPT Int (1):30–36
16. Sellan R, Fabbro M, Burin P (2008) The 300 t EAF meltshop at the new Iskenderun minimill complex. MPT Int (2):52–58
17. Abel M, Hein M (2008) The breakthrough for 320 t tapping weight. MPT Int (4):44
18. Arvedi G, Manini L, Bianchi A et al (2008) Acciaieria Arvedi: a new giant consteel in Europe. In: Proceedings, AISTech conference, Pittsburgh, p 181
19. Evstratov VG, Shakirov ZKh, Zinurov IY et al (2012) Operational experience of 120-t EAF in Asha's metallurgical plant, OAO. Chermetinformation, Bulletin, Iron and Steel industry (12)
20. Manfred X, Fuchs G, Auer W (1999) Electric arc furnace technology beyond the year 2000. MPT Int (1):56–63
21. Abel M, Dorndorf M, Hein M et al (2011) Highly productive electric steelmaking at extra low conversion costs. MPT Int (3):92–96
22. Rummler K, Tunaboylu A, Ertas D (2011) Scrap preheating and continuous charging system for EAF meltshop. MPT Int (5):32–36

Chapter 3
Scrap Melting Process in Liquid Metal

Abstract Analysis of the mechanism of basic processes of melting a scrap in liquid metal including heat transfer from metal to scrap is given. The experimental data obtained by the method of steel samples melting are considered in detail. It is shown how scrap melting time under actual EAF bath conditions can be calculated by using these data. The three most efficient innovations capable of sharp accelerating the melting process are determined. These are: increase in velocity of metal flows at the scrap charging zone due to bath blowing with oxygen tuyeres submerged into the melt to a slag-metal interface; expansion of the volume of the charging zone by means of changing bath configuration as well as using special charging devices; and high-temperature scrap preheating in the shaft with high-power burner devices.

Keywords Melting processes · Carburization · Convection · Heat transfer coefficient · Physical properties of liquid metals · Features of scrap melting in liquid metal · Melting with solidified layer · Melting without solidified layer · Melting of individual piece · Melting of multiple scrap pieces · Experimental data on melting scrap samples · Equivalent scrap · Calculation of melting time · Methods of melting acceleration

3.1 Preliminary Comments

A relatively low rate of melting a scrap in the liquid metal is a principal cause of reduced productivity in conveyor and shaft furnaces in comparison with that of modern traditional EAFs. To select the most effective way for intensification of this process the clear perception in its physical nature as well as bottlenecks associated with the latter is required. The process is characterized by a great complexity. A hot heel, where scrap is melted, always contains a certain amount of carbon. Therefore, not only processes of heat transfer from the melt to scrap but also diffusion processes of saturation of surface scrap layers with carbon can

© The Author(s) 2015
Y.N. Toulouevski and I.Y. Zinurov, *Electric Arc Furnace with Flat Bath*,
SpringerBriefs in Applied Sciences and Technology,
DOI 10.1007/978-3-319-15886-0_3

occur when melting a scrap in the ferricarbonic melt. If any, the scrap melting temperature drops and the process accelerates.

Intensiveness of both heat transfer and carbon diffusion in the liquid metal is mainly determined by hydro-dynamic factors such as: turbulent or laminar fluid motion regime; directions and rates of fluxes flowing pieces of scrap; processes occurring in boundary liquid metal layers contacting solid surfaces; bath stirring intensity. All these factors are examined below, Sect. 3.3.

A complex combination of thermal, diffusive, and hydro-dynamical processes cannot be practically tested by analytical methods of studies. In fact, almost the whole of our knowledge in this field has been gained experimentally under laboratories and plants conditions by using various methods. The results of these experiments are usually represented as relationships between dimensionless similarity criteria (numbers), which are combinations of physical values characterizing the process. An advantage of criterion relationships is that they can be used not only for calculations of isolated cases but for those of the whole group of similar phenomena as well. The similarity theory which is basically the theory of experiment defines a similarity of physical processes as well as rules for selection and usage of similarity criteria. Along with criterion equations, equations with dimensional values can be also used for heat transfer calculations if it is reasonable.

The minimum of information about thermal, diffusive, and hydrodynamic processes in liquid metal which is necessary to substantiate the methods of intensification of melting a scrap in electric arc furnaces with a flat bath is given below. This information is presented in a rudimentary form yet not compromising strict scientific meaning.

3.2 Melting a Scrap with and Without Carbon Diffusion

Melting a scrap in the EAF's freeboard and that in the liquid metal are defined by highly different conditions. In the freeboard, the melting rate is conditioned by processes of heat transfer to scrap from external energy sources such as: electric arc, burners, etc. As mentioned above, the melting rate in the liquid metal containing carbon can depend both on the heat transfer and on saturation of a surface scrap layer with carbon. The carbon is delivered from the melt to the scrap due to the diffusion process. At a low temperature of the melt and high content of the carbon in it, the scrap melting rate can be completely determined by the diffusion process which becomes the basic one.

Under exactly such conditions scrap is melted during the initial stage of the process in oxygen converters. A temperature of hot metal poured into a convertor is lower than the melting point of low-carbon scrap by on average about 200 °C. Melting such a scrap in hot metal would be impossible without carbon diffusion. A melting point of a surface scrap lumps layer when increasing its carbon content becomes lower than a hot metal temperature which rises quickly with oxygen bath blowing. The scrap layer-by-layer is converted into a liquid state and mixes with the liquid bath.

The term "melting" itself does not correspond fully to the nature of such a process inseparably associated with diffusion of carbon. On the other hand, the term "dissolving" is not quite applicable here since dissolving means mixing with the liquid bath through diffusion but without preliminary transition of solid metal into a liquid state. The term "diffusion melting" can be used when describing similar processes. Molecular diffusion is quite slow process in order to be able to serve as a basis for commercial modes of steelmaking. However, in converters this process is many-fold accelerated due to intensive bath stirring with oxygen blowing.

In modern electric arc furnaces making semi-product with subsequent out-of-furnace processing a metal the content of carbon in hot heel is much the same as in scrap. Under such conditions, diffusion melting cannot be noticeably developed.[1] Therefore, only those processes of scrap melting in EAFs, where the rate is practically entirely determined by heat transfer conditions, will be analyzed further.

3.3 Convection as Basic Melting Process Without Contribution of Carbon Diffusion

3.3.1 Definition of Convection; Heat Transfer Coefficient

When melting a scrap in the liquid metal heat is transferred from liquid to scrap due to convection. Convective heat transfer (convection) occurs when a fluid (or gas)[2] flows around a surface of a solid body. Heat can be transferred both from a liquid to a solid surface and in an opposite direction depending on liquid and surface temperatures. Convection is inseparably related to fluid motion. There is no convection in immovable medium.

Convection is a very complicated phenomenon in which, as it will be shown below, conduction heat transfer process takes place as well. Nevertheless, for generalization of experimental data and uniformity of calculation methods, all various convection cases are calculated using the same simple formula (3.1):

$$Q = \alpha (t_L - t_C) \cdot F \tag{3.1}$$

This formula shows that quantity of heat Q transferred from a fluid to a surface of a solid body per unit of time is proportional to the area of this surface F and the difference between the temperature of the fluid t_L and that of the surface t_C. When scrap is melted then $t_L > t_C$. If $t_C > t_L$, heat is transferred from the surface to the fluid.

Coefficient of proportionality α, $W/(m^2\ °C)$, is called coefficient of convection heat transfer. It describes the intensity of the heat transfer process. Formula (3.1) seems very simple only at first glance. In fact, the entire complexity of

---

[1] The exceptions are the furnaces operating with use of hot metal, which are not reviewed here.

[2] At velocities up to 120–150 m/s being very far from the speed of sound, gases behave as noncompressible liquids so that gas and liquid flows are governed by the same laws of motion.

convective heat transfer process and difficulties of its calculation are concealed in the only value, i.e. in coefficient of heat transfer α. This formula is used in calculations of convection processes which are very different by their physical nature. Experimental results are processed in such a way that allows determining the dependence of α on factors which determine conditions of the process run. One of the most important factors is a mode of fluid particles motion.

3.3.2 Two Modes of Fluid Motion

According to their physical nature two modes of fluid motion are distinguished: laminar and turbulent. In laminar mode, separate layers of fluid move parallel to the walls and one another and intermix very slowly. Mixing in this case occurs only due to thermal motion of separate molecules in the direction transversal to the direction of the fluid flow. In turbulent mode, micro- and macro-vortices emerge which move chaotically in all directions and intensely mix the entire fluid. The laminar mode occurs at low fluid velocities while the turbulent mode takes place at high fluid velocities.

As velocity increases transition of laminar mode to turbulent one occurs as soon as the velocity reaches a certain value critical for these conditions. Critical velocity values are different for various fluids and fluxes of different geometry. Virtually uniquely the transition to turbulent mode is determined not by velocity but rather by dimensionless Reynolds criterion:

$$Re = w \cdot d/\nu \tag{3.2}$$

w flux velocity, m/s
d the inner diameter for pipe or specific dimension of bodies flowed around by flux, m
ν kinematic viscosity of fluid, m^2/s

When fluid flows in pipes and channels the laminar mode is observed at the values up to $Re < 2300$. When Re > 2300 the turbulent mode occurs and develops. The fully developed turbulent mode establishes when $Re > 10^4$. Thus, the value Re describes the degree of flux turbulence. An EAF's liquid metal bath, which is stirred both with bubbles of a carbon oxide CO evolving in it and with oxygen jets, is always characterized by a high degree of turbulence. As the criterion Re grows, the coefficient of heat transfer α increases as well.

3.3.3 Boundary Layer

The intensity of heat transfer from the liquid metal to scrap is determined to a large extent by processes operating in thin layers of fluid which contact a solid surface of bodies flowed around. The particles of fluid flow which are in direct contact with the surface of a solid body are absorbed by this surface and quasi-adhere to it. Thus, a very thin immovable layer of the adhered fluid forms on the surface. Since the liquid metal as all fluids has viscosity, forces of viscous friction occur between the layers of fluid which move at different velocities; these forces are proportional to viscosity coefficient ν in Eq. (3.2). These forces lead to slowing down of those layers of fluid which are farther away from the surface when they come in contact with the immovable layer. As a result, a layer of slowed-down fluid of thickness δ, so called hydro-dynamic boundary layer, forms near the surface. In the turbulent flow, at the major portion of thickness the boundary layer is turbulent as well. However, in the immediate vicinity to the surface where velocities of fluid drop down abruptly approaching zero the turbulent mode transitions to the laminar one. This portion of the boundary layer is called laminar or viscous sub-layer, Fig. 3.1. The thickness of this sub-layer δ_1 depends on the degree of turbulence of the main flux which is determined by the criterion Re. As Re grows, the thickness δ_1 decreases.

The heat flux from the fluid to the solid surface should overcome the laminar sub-layer. There is no stirring of the fluid in this sub-layer and heat can spread by only its heat conduction. Therefore, just in the laminar sub-layer despite its small thickness, the major portion of heat resistance to convection heat transfer r_α is concentrated. This resistance is the value reciprocal to α: $r_\alpha = 1/\alpha$. Beyond the laminar sub-layer heat propagation in turbulent flow occurs through stirring of the fluid. This process runs so intensively that the temperature of the fluid virtually does not vary through the cross-section of the flow core since the thermal resistance of this zone is very low. The temperature change occurs mostly within the laminar sub-layer, Fig. 3.1.

In order to increase the coefficient of heat transfer α from the fluid flow to the solid surface, formula (3.1) it is necessary to decrease in some way the thickness of the laminar sub-layer and consequently its thermal resistance. To achieve this value of Re is raised by increasing the velocity of the flow. Increasing of α and intensification of heat transfer can be achieved also by the artificial turbulization of the laminar sub-layer e.g., by jets directed perpendicularly to the surface of the body. Under EAF's conditions, methods for increasing of α due to the improvement of bath stirring intensity are of a great practical importance.

3.3.4 Bath Stirring

The stirring intensity is usually evaluated by the time required for equalizing the bath composition throughout the entire bath volume after introducing a foreign material, a so-called tracer, into any small part of this volume. Under the industrial conditions, the additives of either copper or radioactive isotope of cobalt which are not oxidized in the steel bath are used as a tracer. The so-called bath uniform mixing time is determined by the method of sampling in the course of the heat and analysis of the metal samples from the different points of the bath. The bath

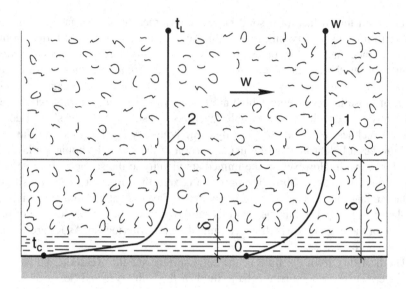

Fig. 3.1 Distributions of velocity (1) and temperature (2) through cross-section of the flow, moving at velocity w along the wall. δ thickness of hydro-dynamic boundary layer, δ_l thickness of laminar sub-layer, t_L temperature of liquid, t_C temperature of the wall

uniform mixing time is characterized by practically identical content of the tracer in all of the samples.

Measuring the stirring intensity in the operating steelmelting units by the above-described method is a very complex and labor-intensive process. All the more attractive appears the option of studying the stirring processes by the methods of cold physical modeling. However, physical modeling gives reliable results only if the definite rules of the similarity theory are complied with. These rules imply quite strict requirements for the selection of both geometric parameters of a model and physical parameters of the liquids simulating metal and slag.

Unfortunately, the rules of the similarity theory are either considered inadequately or even completely ignored in a number of studies. The liquid steel is simulated by water, although the application of water interferes with developing a model in a required scale and does not allow selecting a suitable liquid for the slag simulation [1]. In this book, the results of only those studies, which correspond adequately to the requirements of the similarity theory, are used.

Visual observations show that the melt in EAF is in constant motion. This visible motion of the melt is characterized by circulation in a horizontal plane. The linear velocity of this circulation amounts to 0.3 m/s. However, because of a high volume of the bath, this low velocity corresponds to quite high values of the criterion Re, which confirms the clearly expressed turbulent nature of motion. Besides the horizontal flows, there are vertical flows circulating in the bath, which are characterized by approximately the same velocity and level of turbulence. The ascending flows of the metal contribute to equalizing the temperature over the depth of the bath. All these flows ensure the so-called circulation bath stirring.

The circulation of the melt occurs due to a number of factors. The high-speed jets of oxygen and the floating up bubbles of CO are the key factors. The circulation stirring of the macro-volumes of the bath is of great importance. It ensures the continuous renewal of gas—metal interfaces in the reaction zones of oxygen blowing, the delivery of fresh metal into these zones, and dispersion of the iron oxides formed throughout the entire volume of the bath. This type of stirring plays the dominant role in the processes of bath heating and scrap melting in it.

The second type of stirring occurs due to the high-frequency turbulent pulsations of the micro-volumes of the bath, which are amplified with an increase in the flow velocity and in the criterion Re. This type of stirring is called pulsation stirring. Pulsation stirring ensures the continually renewed close contact of reagents required for chemical reactions to take place in the bath. In particular, it is the pulsation stirring that ultimately ensures the high decarburization rate of the bath required for the current level of intensification of the melting process in the EAFs.

The small-scale turbulent pulsations in the bath which intensity is determined by the criterion Re, are superimposed with the large-scale pulsations of metal caused by the motion of CO bubbles and by the processes in the reaction zones occurring due to penetration of oxygen jets into the bath. These additional pulsations substantially increase the intensity of bath stirring as well as considerably accelerate bath heating and scrap melting. Therefore, circulation and pulsation stirring supplement each other and thus enhance the total effect.

3.3.5 Free and Forced Convection

These two types of convection are distinguished by an energy source which causes the motion of fluid. In the case of forced convection, the motion of fluid is caused by any external agitator. When flowing fluid in pipes, such an agitator is a pump while the bath stirring of EAFs is caused by the jets of oxygen or by the floating-up bubbles of CO. At free convection, the motion of fluid is created by the very heat exchange process and does not require additional energy consumption. In this case, the motion is caused by the difference of densities of heated and cooled fluid layers. Free convection develops the more intensive the greater both the difference between the temperatures of the fluid and the wall and the fluid's volumetric expansion coefficient is.

Near the hot vertical wall, the fluid (e.g. air) is heated, its density is reduced, and the lifting force appears and makes the heated "lighter" air move along the wall upwards. Due to the same reason, near the cold wall the cooled "heavier" air flux appears which moves downwards. In both cases, the convective heat exchange occurs between the wall and air. In the former case, the heat flux is directed from the wall to air which is being heated; in the latter case, the heat flux is directed from air to the wall while air is being cooled.

In comparison with the forced convection, the intensity of the free convection is low. However, it mostly determines heat losses through external surfaces of steelmelting units into environment. Free convection should be taken into consideration when calculating heat balances of: EAFs' bath, heating tunnels of conveyor-type furnaces, and shafts of shaft-type furnaces.

Hot external surfaces release heat not only by free convection to ambient air but also by radiation; in this case the radiation share increases rapidly as the temperature rises. Usually the calculation formulas do not separate these two types of heat losses but consider jointly using of the combined coefficient of heat transfer α_Σ. In the case of the vertical wall, with the temperature t °C of the surface the following formula can be used for approximate calculations of this value:

$$\alpha_\Sigma = 7 + 0.05\,t \quad W/(m^2\ °C) \tag{3.3}$$

The obtained value α_Σ is substituted in formula (3.1) where the hot surface temperature t is taken as t_L, and the ambient air temperature is taken as t_C. Heat transfer by free convection to the ambient air depends on the wall orientation. In comparison with the vertical wall, heat transfer from the wall hot surface of which is directed upwards increases by 20–30 %. If the surface faces downwards heat transfer decreases by the same value. Formula (3.3) can be used if temperatures of the hot wall do not exceed of 400–450 °C. At higher temperatures, heat losses in the environment should be calculated by using formulae of the radiation since an intensity of the radiation increases in proportion to the fourth power of absolute temperature T^4 and in comparison with it the intensity of the free convection becomes insignificant.

The vast numbers of studies are dedicated to the forced convection in various fluids and gases. It is established that in the field of the turbulent developed motion, which is of the greatest interest in this case, the results of all these studies may be summarized by using only three non-dimensional similarity criteria: the Reynolds number $Re = w \cdot d/\nu$, Eq. (3.2); the Nusselt number Nu is expressed as follows:

$$Nu = \alpha \cdot d/\lambda \tag{3.4}$$

and the Prandtl number is given as follows:

$$Pr = \nu/\theta \tag{3.5}$$

λ coefficient of thermal conductivity for a solid body, e.g., a piece of scrap, W/(m °C)

$\theta = \lambda_L/(c_p \cdot \rho)$ coefficient of temperature conductivity for fluid, m^2/s

λ_L coefficient of thermal conductivity for fluid, W/(m °C)

c_p thermal capacity of fluid, W h/(kg °C)
ρ density of fluid, kg/m^3

Applying these three criteria, all the various cases of the forced convection may be described by a general formula (3.6). This is true both for the motion of fluid inside pipes and channels and for external flowing of bodies of various shape and different relative position.

$$Nu = A \cdot Re^x \cdot Pr^y \qquad (3.6)$$

Having determined the value Nu by formula (3.6), the value α is easily found from Eq. (3.4): $\alpha = Nu \cdot \lambda/d$. Coefficient A and powers of x and y change depending on the Re number and the instance of the motion as well. In addition, formula (3.6) is true for either but only one the direction of heat flux, namely, from a fluid to solid surfaces or in the opposite direction. The versatility of formula (3.6) for both the directions of heat flux is achieved by introducing an additive term into it.

In case of the cross flowing of a staggered pipes (cylinders) bundle, Fig. 3.2, with the fluid flow, formula (3.6) takes the form:

$$Nu = 0.41 \cdot Re^{0.60} \cdot Pr^{0.33} \qquad (3.7)$$

In this formula, the mean temperature is taken as a temperature determining physical properties of a fluid; the velocity of the flow at the very narrow section between pipes and diameter of pipes are taken as the velocity and specific dimension determining the Re number, respectively. At such a selection of determining parameters, the heat transfer is virtually independent of a relative distance between pipes.

Fig. 3.2 Profile of fluid motion in the staggered tube bank

This type of convection is of interest in a preliminary estimation of α in the process of melting a scrap comprised by round rods as well since, to a first approximation, it may be considered as a distant analog. Formula (3.7) determines the value of coefficient α for the third row and all the following rows of pipes. For the first row, the obtained value of α should be multiplied by 0.6, and for the second one, that should be multiplied by 0.7. This is explained by the fact that vortices occur behind the first and second rows of pipes which mix intensively the flux, increase the degree of its turbulence and coefficient α, Fig. 3.2. After the first and second rows of pipes an influence of vortices is stabilized and further increase in turbulence is not observed. Identical vortices occur when flowing of scrap lumps with fluxes of the liquid metal as well. This increases the scrap melting rate. Formula (3.7) as well as other formulae of the (3.6)-form might be used in calculations of heat transfer for water, oil, and many different fluids including liquid metals which should be separated out into a special group because of their properties.

3.3.6 Features of Convection in Liquid Metals

In their physical properties liquid metals, particularly the melt of low-carbon steel, differ dramatically from other fluids. In comparison with water the density of liquid steel ρ is much higher, the heat capacity c_p is lower, the coefficient of heat conductivity λ is higher by about 50 times, and the Pr criterion is lower by 60 times, Table 3.1. Such considerable changes in physical properties have required using the Peclet criterion of similarity in formula (3.6) instead of the Re number:

$$Pe = w \cdot d / \theta \tag{3.8}$$

In addition to that, the formula for the case of the cross flowing of a staggered pipes bundle has been derived [2]:

$$Nu = Pe^{0.5} \tag{3.9}$$

or:

$$\alpha d / \lambda = (wd/\theta)^{0.5} \tag{3.9'}$$

The expression for the mean value of α in the bundle after the second row of pipes follows from formula (3.9'):

Table 3.1 Physical properties of molten low-carbon steel over the range of temperatures from 1570 to 1590 °C in comparison with those of water

	ρ (kg/m^3)	c_p (Wh/(kg °C))	λ (W/(m °C))	θ (m^2/s)	v (m^2/s)	Pr
Steel	7000	0.24	3.2	5.3×10^{-6}	0.65×10^{-6}	0.12
Water	998	1.16	0.597	0.143×10^{-6}	1.00×10^{-6}	7.0

$$\alpha = \frac{\lambda}{d}\left(\frac{w \cdot d}{\theta}\right)^{0.5} \tag{3.10}$$

w flow velocity at narrow sections of the bundle, m/s
d external diameter of pipes, m
θ coefficient of temperature conductivity, m²/s, see Sect. 3.3.5

As a rule, studies of convective heat transfer in liquid metals were carried out with the use of low-melting metals. In particular, formula (3.9) has been obtained from experiments using mercury and sodium which a melting point is as low as 98.7 °C [2]. In similar studies, there was no formation of solidified metal layers on a surface of bodies immersed into the melt. When melting a scrap in the liquid steel the solidification phenomenon has a strong effect on the melting time. Therefore, it becomes necessary to consider particularly the results from not numerous experiments with liquid steel where the solidification has taken place. Just these studies of convection are of the greatest interest for analyzing the scrap melting process in EAFs with a flat bath.

3.4 Melting in Liquid Steel; Experimental Data

3.4.1 Melting of Individual Piece of Scrap with Solidified Layer

One of the basic and most frequently used experimental methods of study of melting processes is a method of melting samples immersed into the melt. The case is known when this method has been also tried under industrial conditions, but on the open-hearth furnace. Unfortunately, with regard to EAF, only the laboratory experimental data are available. Let us review the above mentioned case, since many processes including melting are similar for the hearth furnaces.

In the work [3], the slabs with the thickness of 200 mm were immersed in the bath of 650-t open-hearth furnace in the zone where the metal temperature was continuously controlled by special thermocouple. Experiments were carried out during the final stage of the heat. During this stage the metal was heated from 1570 to 1620 °C with oxygen blowing into the bath with small specific intensity 0.10–0.12 m³/(t min). After immersion of the slab, the thermocouple registered the local reduction in metal temperature by 10–12 °C, which lasted for 5–10 min. Then temperature dropping caused by cooling effect of the melting slab stopped, and the temperature started to rise at the same rate as it took place prior to slab immersion. It was assumed that this moment corresponds to completion of slab melting. This way the duration and average linear melting rate of the slab was determined, and the average linear melting rate amounted from 12.3 to 15.8 mm/min. Using this rate value and invoking several simplifying assumptions, the coefficient α of convective heat transfer from liquid metal to the slab surface was found. The average value of α was equal to 27, 800 W/(m² °C), which is consistent with the results of calculations in Sect. 3.4.3, see below.

However, using this experimental technique, it was impossible to extract the slab from the bath during its melting. Inevitable under the conditions of the experiment solidification of metal on the slab surface was not controlled and accounted for. Therefore, the actual rate of the change of the slab thickness per time remained unknown, and calculation of α using average melting rate could produce approximate results only.

The results of study of melting processes by method of melting samples obtained under laboratory conditions in 2004–2005 at McMaster University, Hamilton, Canada [4][3] are of great practical interest. Cylindrical samples made of low-carbon steel were immersed in the crucible of induction furnace containing liquid metal. The chemical composition of the metal was close to the composition of the samples. The diameter and the initial temperature of samples, the duration of their immersion into metal were varied. At fixed moments of time, the samples were taken out from the crucible, cooled in water, and weighed. The radius of the immersed part of the samples R was calculated using change in mass. Based on the data from these experiments, the curves of sample melting were plotted.

There are two possible alternatives of melting: with or without solidifying of metal layer on a sample. Realization of one or the other of these variants depends on the temperature of the metal and the initial temperature of the sample. These temperatures determine the value of the three different by their physical nature heat flow taking place in the process of melting. First of them q_1 is the convective heat flow from liquid metal to sample surface or to solidified layer, Fig. 3.3. This flow is determined by the equation:

$$q_1 = \alpha(t_L - t_C), \ W/m^2. \tag{3.11}$$

α   coefficient of heat transfer from liquid metal to solid surface of the layer or of the sample W/(m² °C)

t_L and t_C temperatures of liquid metal and solid surface, respectively, °C, Fig. 3.3

The heat flow q_2 occurs due to thermal conductivity of the layer and the sample. It is directed from the surface of the sample toward its axis; it is consumed to increase the temperature of the sample, and is determined by the equation:

$$q_2 = \lambda \cdot \frac{dT}{dR} \tag{3.12}$$

λ   coefficient of thermal conductivity of layer and sample, W/(m² °C)
$\frac{dT}{dR}$ temperature gradient on sample surface

----

[3] In this chapter, the data from electronic publications of the authors of work [4] have also been used.

Fig. 3.3 Schematic diagram of temperature profile of steel bath and sample immersed into it. *1* liquid metal surface, *2* solidified layer, *3* steel sample

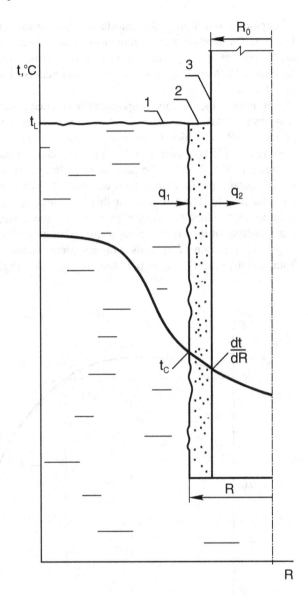

The heat flow q_3 is related to the conversion of metal from liquid into solid state or vice versa at constant temperature t_C. This flow is determined by the equation:

$$q_3 = H \cdot \rho \cdot v \tag{3.13}$$

H 75 W h/kg—latent heat of melting (solidifying) of low-carbon steel
ρ 7900 kg/m^3—density of solid metal
v speed of interface migration between solid and liquid metal, m/h

In case of solidifying, this interface moves to the left, and H has a plus sign, Fig. 3.3. The thickness of solidified layer increases, and heat is emitted. During the process of melting, the interface travels to the right, the thickness of layer decreases, H changes its sign to minus, and heat is absorbed. For the cylindrical sample, $v = \frac{dR}{d\tau}$.

Let us examine the most important for practice case: melting with solidifying. It is always takes place when cold scrap is immersed in metal with temperature of 1550–1650 °C which is common for EAFs and converters. A typical curve of melting cold (25 °C) sample of 25.4 mm diameter in metal at temperature 1650 °C is shown in Fig. 3.4 [4]. The process of melting with solidifying consists of two stages. During the first stage with duration of τ_C, metal layer is being solidified onto the sample. The thickness of this layer first increases, reaches maximum at point m, after which the layer begins to melt and disappears at point n, Fig. 3.4. After melting of the layer is completed, melting of the sample starts. This second stage with duration of τ_L ends with complete melting of the sample at point x. Total melting time $\tau = \tau_C + \tau_L$ in this case is 37.5 s, Fig. 3.4.

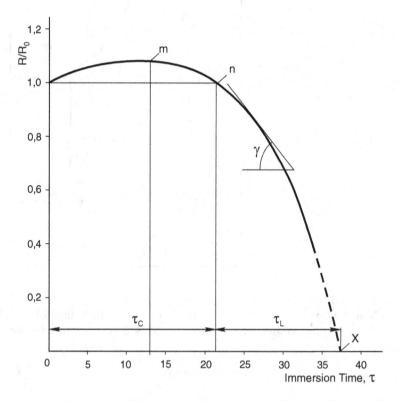

Fig. 3.4 Dependence of relative sample radius, R/R_0, on increase in time of immersion into the melt, τ [4]. τ_C period of solidification of metal onto sample and melting of solidified layer, τ_L period of melting a sample itself, d 25.4 mm; t_L 1650 °C

To better understand mechanism of melting of scrap in liquid metal as well as to determine later on the effectiveness of various methods of intensification of this process, a detailed analysis of the shape of the melting curve, Fig. 3.4, is needed and, in particular, of presence of maximum of this curve at point m. Due to quite high value of α in case of heat transfer in liquid metal, the temperature of sample surface right after immersion increases very rapidly to melting (solidifying) point of the metal, t_C. The heat flow to the sample surface q_1 is stabilized at the value determined by Eq. (3.11). Since the values of α, t_L, and t_C do not change further, the flow q_1 remains constant during the whole process of sample melting. At the same time, a thin metal layer contacting the sample surface cools down and solidifies at the same temperature t_C.[4] Following the first layer, the second, third, and so on layers also solidify. The thickness of the layer solidified on the sample increases. The temperature on its surface stays constant and is equal to t_C.

The heat from the solidified layer transfers to the sample due to thermal conductivity (flow q_2), Fig. 3.3. At the first moment after immersion of the sample, both gradient of temperature on the sample surface $\frac{dt}{dR}$ and flow q_2 have maximum values, Eq. (3.12). Then, as the temperature of sample rises, the gradient $\frac{dt}{dR}$ and the flow q_2 are reduced. However, due to high thermal conductivity of the sample, the condition $q_2 > q_1$ remains unchanged for a certain period of time.

During this period, the intensity of metal cooling by the sample is gradually reduced. The thickness of the solidified layer continues to increase; but the rate of this increase gradually drops (melting curve in the interval from the beginning of the process to point m, Fig. 3.4). The heat balance of the metal—sample system, for which heat input is required to be equal to heat output, is maintained due to latent heat H of metal converting from the liquid to the solid state. The evolution of this heat produces the heat flow q_3, Eq. (3.13) which adds to the flow q_1 and keeps the balance:

$$q_1 + q_3 = q_2 \qquad (3.14)$$

Approximately 13 s after immersion (point m), the sample temperature rises to such an extent that the intensity of liquid metal cooling becomes not high enough for the process of solidification of the layer to continue. The further growth of the layer thickness ceases and the evolving of latent heat H stops. At the point m, $q_3 = 0$ and $q_1 = q_2$. Even greater increase of the sample temperature results in the further reduction of the flow q_2 which becomes smaller than q_1, $q_2 < q_1$. At that, the layer solidification is replaced by its melting. The heat flow q_3 changes its sign. Now this heat is not evolving in the layer, but is consumed by melting of the layer, which ends at the point n. Then the melting of the sample itself starts, which

---

[4] In iron-carbon melts, this transition starts at the liquidus temperature and ends at the lower solidus temperature. For the low-carbon steels, as in our case, the temperature difference between these two points can be neglected. Therefore, from now forth we will consider that the metal solidification and melting processes occur at the same constant temperature, i.e. melting point $t_C \cong 1530 \,°C$.

ends at the point x, Fig. 3.4. The same balance equation corresponds to the melting curve in the interval from point m to point x:

$$q_1 = q_3 + q_2 \tag{3.15}$$

At the end of melting, the average-mass temperature of the sample becomes approximately equal to t_C, and the value of q_2 can be disregarded as compared to q_3. At this last interval, the melting curve practically coincides with a straight line. At this, the melting rate, which is characterized by the tangent of the angle γ, increases up to its maximum value, since practically the entire flow q_1 is used for melting of the sample. Extrapolation of the straight line allows to determine the position of the point X with sufficiently high accuracy, Fig. 3.4.

3.4.2 Melting Without Solidified Layer

Complete elimination or even partial shortening of the period of the solidified layer existence considerably decreases duration of scrap melting. Let us examine the conditions at which scrap can be melted in liquid metal without solidifying. As the analysis of the melting curve demonstrates, solidifying occurs only under conditions of $q_2 > q_1$. Therefore, in order to eliminate solidification, it is necessary to increase the heat flow to the scrap surface q_1 and to decrease the heat flow q_2.

According to Eq. (3.11), the heat flow q_1 rises with an increase of heat transfer coefficient α as well as of temperature of liquid metal t_L. The possibilities of increasing α are discussed below in Sect. 3.5. In regard to the temperature t_L, the formation of the solidified layer is completely eliminated when this temperature is high enough. However, under actual operating conditions of steel production, this temperature is determined by process conditions, and its substantial increase does not seem possible.

The most effective method of elimination of solidification is decreasing the heat flow q_2 by scrap preheating. As analysis of data published in the work [4] demonstrates, the duration of the period of solidified layer existence shortens as preheating temperature t_S increases, and at $t_S \cong 890$ °C or higher solidification does not takes place anymore, Fig. 3.5.

The calculations using the results obtained by the method of sample melting in the works [4–6] have shown that the ratio of duration of existence of solidified layer τ_C to total duration of sample melting τ does not depend, over rather wide range, on the diameter of the sample and is equal to 0.6, Fig. 3.6. This self-similarity of the process is noted here for the first time. It can be used for the analysis of the effect of scrap preheating temperature on duration of melting of not only an individual piece, but also on duration of simultaneous melting of multiple various scrap pieces.

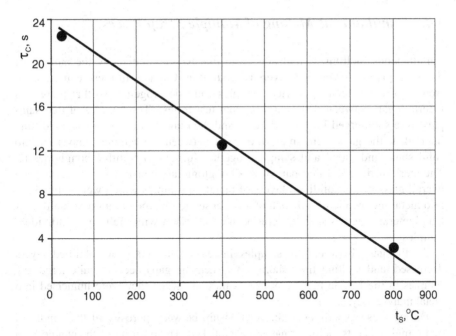

Fig. 3.5 Dependence of existing time of solidified layer τ_C on sample preheating temperature t_S

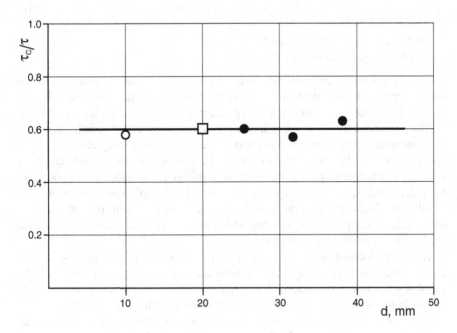

Fig. 3.6 Dependence of ratio, τ_C/τ on sample diameter, d. ● [4], ○ [5], □ [6]

3.4.3 Simultaneous Melting of Multiple Scrap Pieces

Simultaneous melting of multiple scrap pieces was studied by the authors of the work [4] at McMaster University, Hamilton, Canada. In these experiments, from two to nineteen cylindrical samples were tied together so that there was a certain gap between them varying from test to test. The bundles of tied samples were submerged into liquid metal and then taken out of it at specified time interval. If the gaps were small, the metal between the samples converted into solid state, and metal and samples together formed monolithic conglomerate. The mass of the solidified part of this conglomerate considerably exceeded the overall mass of the solidified layers of metal on samples which were submerged into metal independently of each other. Consequently, the duration of melting of conglomerate increased considerably in comparison with melting of individual samples.

As the gaps between the samples increased, the mass of solidified layers decreased, and melting time shortened. When the gaps became quite large, the samples in the bundle melted as fast as they did when they were immersed into metal individually.

During these experiments, the relationship between porosity of the bundle of tied samples and its melting time was established. The porosity of the bundle P, % is determined by the equation:

$$P = (1 - \upsilon_{st}/\upsilon_{\Sigma}) \cdot 100 \text{ \%} \qquad (3.16)$$

υ_{st} volume occupied by samples, m^3
υ_{Σ} total volume of the bundle, m^3

As the porosity increased, the duration of melting of the bundles decreased and asymptotically approached the duration of melting of individual samples. At $P \geq 96$ %, the bundles melted practically as fast as individual samples, Fig. 3.7 [4].

The actual conditions of scrap melting in liquid metal differ significantly from the conditions in experiments described above. Continuous scrap charging in the furnaces with flat bath is carried out into quite limited zone of liquid bath. In the furnaces of medium capacity, the volume of this zone increases in the course of the process but is, on the average, no more than 2.5–3.0 m^3. Some pieces of scrap in this zone locate on the bottom and onto each other. At high rate of charging, the surface of scrap pieces contacting with metal can be significantly smaller than total surface area of individual pieces.

Randomly located in the charging zone, pieces of scrap form a 3-dimensional lattice which has certain hydraulic resistance, which reduces the speed of the metal flows between the pieces. The metal temperature in the zone can drop significantly in comparison with the temperature outside of the zone. All these factors decrease heat flow from metal to scrap and slow down the melting. On the other hand, vortices are formed in the flows around the scrap pieces which increase heat transfer intensity in some degree.

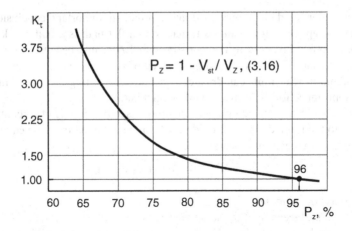

Fig. 3.7 Dependence of scrap portion melting time τ on charging zone porosity P_Z. Coefficient K_τ is equal to ratio of τ to melting time of individual scrap piece

Because there are no other more reliable data, it can be considered, by analogy with melting of the bundles, that Eq. (3.16), Fig. 3.7, characterizes, with acceptable accuracy for practical conclusions, an effect of the charging zone porosity P_Z on melting time of a scrap portion under actual conditions of the furnace bath. As per this equation, the maximal possible melting rate is achieved at the $P \geq 96\ \%$. However, such a level of the porosity requires too large volumes of the charging zone and so cannot be realized in practice.

Scrap charging evenly distributed throughout the furnace bath would be ideal. In this case, the charging zone volume is equal to that of the entire metal furnace bath. Using Fig. 3.7 let us determine the total melting time of the entire scrap in a furnace of 185 t capacity for this ideal charging method. With the capacity of 185 t and tapping weight of 130 t, the liquid metal weight on average in the course of the heat amounts to 120 t,[5] and its volume is: $120/7.0 = 17.1\ \mathrm{m}^3$. In the present case, this volume is the mean volume of the charging zone V_Z. Let us preset the value of the zone porosity P_Z equal to 0.90 in which $K_\tau = 1.1$ (Fig. 3.7) the scrap portion melting time is longer than that of individual pieces by a factor of 1.1 and amounts to: $31.5 \times 1.1 = 34.6\ \mathrm{s}$ or 0.58 min. Knowing P_Z and V_Z, and using formula (3.16), see Fig. 3.7, let us find the total volume of pieces in the scrap portion V_{st}: $V_{st} = V_Z\,(1 - P_Z)$; $V_{st} = 17.1(1 - 0.90) = 1.71\ \mathrm{m}^3$. The portion mass of $M_{st} = 1.71 \times 7.9 = 13.5\ \mathrm{t}$.[6] This portion of scrap will be melted for 0.58 min, and, consequently, the entire scrap in $0.58 \times 143/13.5 = 6.1$ min.

Although calculated estimation of the duration of melting down of 143 tons of scrap in the furnace of 185 t capacity equal to approximately 6 min is probably

---

[5] The weight of hot heel is 55 t. Half a tapping weight amounts to 65 t and so: $55 + 65 = 120$ t.

[6] Density of solid steel $\rho = 7.9$ t/m^3.

somewhat understated, it allows to draw a principal important conclusion. An insufficient scrap melting rate in the furnaces with flat bath is a bottleneck which limits their competitiveness in comparison with EAFs just because the scrap is continuously charged into the liquid metal zone of too small volume. An increase in this volume in real limits can drastically accelerate melting of scrap. This is of primary importance for the furnaces with the flat bath.

The total formula for the duration of melting of scrap τ in case of its continuous charging into the limited zone of liquid metal can be obtained from the equation of heat balance of the melting process:

$$\alpha(t_L - t_C)F \ \tau \ = M[H + c_p(t_C - t_S)] \tag{3.17}$$

M   mass of scrap, kg
F   total surface area of the scrap pieces, m^2
H, c_p latent heat of melting, W h/kg, and heat capacity of a solid piece, W h/(kg °C)
α coefficient of heat transfer, $W/(m^2 \ °C)$
t_L, t_C, t_S temperatures, °C, of liquid metal, of the surface of melting scrap piece, and of scrap preheating, correspondingly
τ melting time, h

The left side of Eq. (3.17) is a quantity of heat obtained by scrap during time τ, and the right side is heat consumption for melting at temperature t_C. The quantity of heat evolved during solidification of metal layer on scrap pieces is equal to the quantity of heat which is later on consumed for the melting of the solidified metal. Therefore, neither of these heat quantities is considered in Eq. (3.17).

If α is known, the formula for τ follows from Eq. (3.17):

$$\tau = \frac{M[H + c_p(t_C - t_S)]}{F \ \cdot \ \alpha(t_L - t_C)}. \tag{3.18}$$

3.5 How to Accelerate Melting; Analysis of Options

3.5.1 Increasing Heat Transfer Coefficient α

Scrap melting time decreases directly proportional to an increase in the coefficient of convective heat transfer α, Eq. (3.18). Therefore, increasing α is quite an important task. Equation (3.10) describing forced convection in case of the cross flowing of liquid steel through staggered cylinder bank shows that the coefficient α increases directly proportional to an increase of the speed of the flow $w^{0.5}$ and to decrease of $d^{0.5}$, Sect. 3.3.6.

The most affordable and effective method of increasing the speed of the metal flows around the scrap pieces is increasing the intensity of stirring of the steel bath in the charging zone and its vicinity. Unfortunately, very little data is available linking α and the intensity of stirring of the steel bath. In the work [5] which addresses this problem and is of practical interest, the method of melting of samples similar to that in the work [4] has been used. Cylindrical samples were immersed for several seconds into a crucible of induction furnace with liquid steel at temperatures from 1580 to 1800 °C. Experiments were carried out with three different intensities of bath stirring, that is: in the power-off furnace, when heat transfer was realized by natural convection only, as in the work [4]; in the power-on furnace with induction stirring; and in the power-on furnace with additional stirring by argon blown from the top.

In comparison with natural convection, relatively weak induction stirring increased coefficient α by a factor of 1.3 and along with additional argon stirring by a factor of 1.7, regardless of the metal temperature. This purely qualitative evaluation of stirring intensity does not allow obtaining the quantitative dependences. It is obvious, however, that by improving the bath blowing methods it is possible to considerably increase intensity of heat transfer from metal to scrap and to sharply reduce scrap melting time. Unfortunately, at present, data on the actual speed of metal flows in the scrap charging zone of the furnaces with flat bath are not available at all.

The second practically important option for increasing α is decreasing the thickness of scrap pieces. There is an inversely proportional relationship between α and $d^{0.5}$, and not only in case of the forced convection as per Eq. (3.10), but in case of natural convection as well. This is confirmed by the data of the work [4] in which cylindrical samples were melted in liquid metal in the turned off induction furnace. The coefficients α were not being determined in this study. However, the melting curves given in this work allow to determine melting time of the samples with various diameters and, using the values of melting time obtained, to calculate corresponding values of α in accordance with Eq. (3.19):

$$\alpha = \frac{m\left[H + c_p(t_C - t_S)\right]}{F \cdot \tau(t_L - t_C)} \tag{3.19}$$

Initial data and the results of calculations are given in Table 3.2. The considered relationship is confirmed by practically identical (within the limits of experimental accuracy) values of $(d_1/d_2)^{0.5}$ and α_1/α_2. Since the dependence of $d^{0.5}$ on α is common for both forced and natural convection in a quite wide range of variation of the cylinder samples, it can be assumed that this dependence is also valid for scrap pieces of any shape under condition that the defining dimension d characterizes the ratio m/F.

Table 3.2 Experimental values of τ [4] and results of calculations of α as per formula (3.19)

t_S (°C)	d (mm)	τ (s)	α (W/(m² °C))	$(d_1/d_2)^{0.5}$	α_1/α_2
25	$d_1 = 38.1$	$\tau_1 = 75.0$	$\alpha_1 = 13{,}930$	1.22	1.3
	$d_2 = 25.4$	$\tau_2 = 37.5$	$\alpha_2 = 18{,}560$		
800	$d_2 = 25.4$	$\tau_2 = 22.5$	$\alpha_2 = 17{,}800$	1.22	1.3

Therefore, reducing scrap size during its preparation for melting in the flat bath furnaces shortens melting time not only due to decrease in the mass of the pieces, as it was shown earlier. The additional effect can be ensured by an increase of the coefficient α which characterizes the intensity of heat transfer. The combined effect of decreasing the thickness of pieces can considerably increase furnace productivity. However, the fine scrap occupies larger volume. Therefore, in order to completely realize the advantages of the use of a finer scrap, quite large volume of charging zone is required. Let us examine the possibilities of increasing this volume.

3.5.2 Increasing Volume of Charging Zone

The depth of steel bath in the charging zone can be increased by raising the mass of hot heel. Tenova was the first company which used this approach and increased the share of the hot heel in the total mass of liquid metal from 20–25 to 50–55 % in the Consteel furnaces. This significantly increased the rate of melting. The mass of hot heel in the shaft furnaces with flat bath must be also increased.

A further increase of the volume of the scrap charging zone requires increasing the surface area of the zone which is at present limited by the electrodes positioned at the centre of the furnace as well as by the devices for scrap discharging into the bath. It is necessary to improve the design of these devices and to change both the position of electrodes and the shape of the freeboard of the furnaces.

The existing circular shape of the freeboard and the symmetrical positioning of electrodes in the freeboard are the most reasonable in case of melting the scrap which fills the entire free volume of the furnace. In the furnaces with continuous scrap charging, scrap melting zone is adjacent to the bottom bank and occupies a relatively small fraction of the surface area of the bath. In case of this asymmetric position of the zone, increasing its area requires asymmetric positioning of the electrodes similar to that of the Fuchs' shaft furnaces with the shaft located above the furnace roof, Chap. 2, Sect. 2.3.1, Fig. 2.6. Furthermore, in this case, the oval shape of the freeboard might prove to be more suitable than the circular shape.

Special attention must be given to developing more advanced devices for discharging scrap into the bath. This device must ensure the uniform distribution of scrap throughout the entire increased area of the charging zone. Schematic diagram of such a device is examined in Chap. 6.

3.5.3 Scrap Preheating

The analysis of the results of experiments with melting samples obtained in the works [4–6] lead to the following conclusions, Sect. 3.4.2. In case of scrap preheating to 890 °C or higher, the scrap continuously charged into the bath will be melted without metal solidification on its surface, Fig. 3.5. For cold scrap melting, regardless of the dimensions of the scrap pieces, the duration of a solidified layer existence comprises about 60 % of total melting time. This was determined for the samples with diameters from 10 to 40 mm, Sect. 3.4.2, Fig. 3.6. Such a result can be assumed for regular furnace scrap as well. The statement above leads to the conclusion that scrap preheating to the average-mass temperature of 800–900 °C sharply shorten the scrap melting time.

References

1. Markov BL (1975) Methods of open-hearth bath blowing. Metallurgy, Moscow
2. Isachenko VP, Osipova VA, Sukomel AS (1981) Heat transfer. Energoizdat, Moscow
3. Medzhibozhski MY, Shemyakin AV, Lykin AA et al (1972) Experimental determination of heat transfer coefficient for boiling steel melting bath. Izwestiya VUZov, Ferrous Metall 10:52–55
4. Li J, Brooks GA, Provatas N (2004) Phase-field modeling of steel scrap melting in a liquid steel bath. In: AISTech conference, vol 1. pp 833–843
5. Fleisher AG, Kuzmin AL (1982) Effect of temperature of the melt on heat transfer to the surface of an immersed melting body. Izwestiya VUZov, Ferrous Metall 4:40–43
6. Glinkov GM, Bakst VY, Megzhibozhski MY et al. (1972) Melting a cold steel scrap in the overheated ferricarbonic melt. Izwestiya VUZov, Ferrous Metall 3:62–64

Chapter 4
Methods of Realization of High-Temperature Scrap Preheating

Abstract Specifics of furnace scrap facilitating and hindering intensive heating of it up to high temperatures. A shaft furnace as a steel melting aggregate largely satisfying requirements of high-temperature scrap preheating and continuous melting it in liquid metal as well. Inefficiency of a flow of off-gases leaving a furnace to provide such heating because of insufficient heat power of this flow. Inapplicability of well-known oxy-gas burners for this purpose. The system of high-temperature scrap preheating in a furnace shaft with high-power oxy-gas burner devices of recirculation-type is considered. Designed parameters of the system are given. This system has been developed by the authors. It is characterized by a low temperature of flames. Off-gas energy is used in the system for decomposition of dioxins contained in combustion products leaving the shaft and also can be used for other purposes not attached to the scrap melting process, for example, for production of steam with technological and energy parameters.

Keywords Specifics of furnace scrap · Possibility of heating without rate limitation · Oxidizing and welding of pieces · High-temperature scrap preheating · Optimum heating conditions · Heating with off-gas flow · Low heat power of the flow · Heating with gas fuel · Requirements to burner devices · Temperature of the flame · The system of scrap preheating · High-power recirculation burner devices · Dioxins and carbon monoxides in combustion products · Methods of environment protection

4.1 Specifics of Furnace Scrap Associated with Its Heating

In EAFs, as a rule, the cheapest light scrap is used. It has low volume density of approximately 0.65–0.75 t/m³. Such a scrap consists mostly of pieces with relatively small mass and thickness. The length and shape of these pieces vary widely. The denser, cleaner and more expensive scrap is used in converters which are not suitable for melting light scrap. Intent of metallurgists to use cheap scrap in EAF

© The Author(s) 2015   69
Y.N. Toulouevski and I.Y. Zinurov, *Electric Arc Furnace with Flat Bath*,
SpringerBriefs in Applied Sciences and Technology,
DOI 10.1007/978-3-319-15886-0_4

is determined by the fact that cost of scrap accounts to approximately 70 % of total cost per heat of materials, energy and personnel.

Depending on the source of scrap supply and the method of its preparation for melting the thickness of scrap lumps varies from a few millimeters (sheet bushelling) to 100–150 mm. Internal thermal resistance of such pieces is so low that each single lump as well as the entire scrap in the shaft can be preheated at any practically achievable rate. The temperature difference between the surface of pieces and their centre remains negligible. This is not true for the relatively large bales. The bales are heated through quite slowly, and therefore their use should be avoided. The low thermal resistance of scrap pieces is of importance for furnaces with flat bath since that is the condition required for rapid scrap melting down in the liquid metal.

Though the scrap for EAF is preselected, it always contains some amounts of rubber, plastics and other flammable organic materials including oil. The chips from metal cutting machines are especially contaminated with oil. The chips are produced in large amounts and require utilization. Oil and other flammable contaminants present in the scrap emit a lot of heat while burning out. This causes quite undesirable consequences. Even when moderate-temperature gas is used for preheating of scrap, pockets of burning and melting of small fractions can be formed in the heated layer. When this occurs, the separate scrap pieces are welded together forming so called "bridges" which obstruct normal discharging of preheated scrap from the shaft into the furnace. Because of this preheating of metal chips is usually avoided. Thus, the specifics of the steel scrap utilized in EAF create certain difficulties for its preheating, especially for the high-temperature preheating. These specifics as well as problems associated with the environment protection against dioxins should be taken into account when selecting a furnace with the flat bath which is most suitable to realize such scrap preheating, Chap. 2, Sect. 2.3.3.

4.2 Selection of the Furnace Type

4.2.1 Conveyer or Shaft Furnace

Experimental data and calculations given in Chap. 3 show that high-temperature scrap preheating is the effective means of the sharp shortening of duration of melting of scrap in liquid metal. Earlier, when tap-to-tap time in EAF was 1.0–1.5 h, it seemed possible to carry out such heating in the furnace freeboard by means of high power rotary oxy-gas burners [1]. At present, with tap-to-tap time of 30–35 min, such possibility does not exist. High-temperature scrap preheating can be carried out in the external devices only, such as a shaft in the shaft furnaces or a conveyor in the Consteel furnaces. At the present time, these devices are used for scrap preheating by the off-gases. However, actually achieved average-mass temperatures of scrap preheating do not exceed 450 °C in the shaft furnaces and 250 °C in the Consteel furnaces, Chap. 2, Sects. 2.2 and 2.3.

The scrap heating system in the Consteel furnaces is barely suitable for a substantial increase in scrap temperature. This increase is hindered by the preheating method itself. Scrap is placed on the mobile chutes of the conveyor and is heated in the tunnel of the furnace from the top. Such configuration does not allow the heating gases to infiltrate through the entire thickness of the scrap layer (as it takes place in the shaft furnaces), which considerably worsens heat transfer conditions. In addition, achieving high average-mass scrap temperatures is limited by inadmissible overheating of its upper layer and is way more problematic than in the shaft furnaces.

Overheating of the upper layer leads to combustion and surface melting, and small fractions of scrap are welded together forming large conglomerates, which disrupts the work of conveyor. In the shaft furnaces, overheating of separate small zones of scrap does not hinder the normal course of the heat. That is why the possible average-mass scrap temperature is considerably higher in the shaft furnaces, which makes them more preferable.

4.2.2 Shaft Furnaces with Scrap Retaining Fingers (QUANTUM-Type) or Furnaces with Scrap Pushers (COSS-Type)

Design and operation of these furnaces are described in Chap. 2, Sects. 2.2 and 2.3. In comparison with the Quantum furnaces, the COSS-type furnaces with the shaft located at the side are more suitable for high temperature scrap heating since they have the following advantages.

In the COSS furnaces, the pushers can charge scrap into the bath by relatively small portions which reached the maximum heating temperature. This is the most optimal charging regime. It is closely comparable to the continuous conveyor charging and fully corresponds to the operating mode of the furnace with flat bath. In the Quantum furnaces, only one portion of scrap at a time can be heated on the shaft fingers. In order to achieve the maximum productivity of the 100-t furnace, the number of the portions charged into the bath should not be more than three [2]. This three-basket charging does not allow to fully realize the advantages of the continuous scrap melting in the flat bath. Furthermore, the average-mass temperature of each portion of scrap in case of such charging is always lower than maximum temperature achievable in case of continuous charging.

The height of the layer of scrap in the shaft of the COSS furnaces exceeds the height of the layer on the fingers. Accordingly, the passage of the heating gases travelling through the scrap is longer, and their heat efficiency coefficient is higher. In the COSS furnaces, this coefficient reaches the value of 0.7. Taking into consideration all the factors discussed above the considerably greater effectiveness of scrap preheating can be expected in the COSS furnaces.

The water-cooled mobile fingers in the shafts of the Quantum furnaces are directly hit by the pieces of scrap falling from the top at least three times during the heat. As a result, despite significant improvements, this structural element remains the weak spot of the system. Obviously, in order to avoid damaging of the fingers, stricter requirements must be imposed on scrap preparation for the Quantum furnaces than for the COSS furnaces.

Another advantage of the COSS furnaces is the possibility to charge scrap into the shaft without dust-gas emissions. Entrapping of these emissions leads to a sharp increase of the volume of gases requiring cleaning and of corresponding costs, Chap. 2, Sect. 2.3.3.

4.3 Selection of the Energy Source

4.3.1 Off-Gases; Insufficient Heat Power of the Flow

As already mentioned in Sect. 2.3, Chap. 2, one of the main causes of relatively low temperature of scrap preheating in the shaft furnaces is insufficient heat power of off-gases flow P_{gas}, kWh/(t min). This key parameter requires detailed quantitative analysis. Unfortunately, there is no data available for P_{gas}. The specific values of heat losses with off-gases Q_{loss}, kWh/t of liquid steel, are given in many works. According to the data from different authors, Q_{loss} ranges from 80 to 220 kWh/t. However, the value of Q_{loss} cannot substitute for P_{gas}, since it characterizes quantity of heat rather than power of heat flow. Using the value of Q_{loss}, it is impossible to determine average-mass scrap temperature t_S which can be obtained in case of scrap heating by off-gases in a given heating time τ. This temperature depends not only on τ, but also on the coefficient of heat transfer from gases to scrap. When τ is very small (close to zero), the scrap remains cold at any value of heat power of the off-gases flow.

In order to calculate achievable temperature t_S, let us examine operation of the COSS-type shaft furnace under the following conditions. The furnace operates with the closed slag door, so that there is practically no air infiltration into the freeboard. The sources of carbon are the lump coke charged through the opening in the roof of the furnace as well as the coke powder injected into the bath. The carbon content in the coke is 80 %.

Carbon contained in scrap is not considered, since its concentration is small (≤ 0.20 %) and does not differ substantially from carbon concentration in liquid steel.

The furnace operates with the specific intensity of oxygen blowing in the bath $J = 0.9$ м^3/(t min) of liquid steel, which is close to the maximum values achievable in practice. All CO evolving from the bath is sucked into the shaft and only there burns to form CO_2. Thus, it is assumed that the off-gases consist of CO only, and that all the chemical energy from CO combustion evolves in the shaft. It is also assumed that duration of scrap preheating by off-gases is 18 min and total tap-to-tap time is 30–32 min.

The assumed conditions of furnace operation are the most favourable for achieving the maximum scrap temperatures t_S. Calculation of t_S under these conditions are given below.

4.3.1.1 Calculation

In case of oxygen blowing in the bath with high content of carbon, the intensity of CO evolution is determined by the equation: $V_{CO} = 2$ kJ, $m^3/(t\ min)$ [1], where $K = 0.7$ is the coefficient accounting for the fact that O_2 is partially spent on oxidation of Si, Mn, and Fe. Besides, a certain amount of O_2 is not absorbed by the bath. The coefficient 2 corresponds to the reaction: $2C + O_2 = 2CO$. $V_{CO} = 2 \times 0.7 \times 0.9 = 1.26\ m^3/(t\ min)$. All volumetric flow rates are given in standard m^3 (s.t.p.).

The coke consumption G $kg/(t\ min)$ corresponds to the flow V_{CO} equal to $1.26\ m^3/(t\ min)$. The mass of carbon in the coke is 0.8 G. The mass of carbon monoxide is $G_{CO} = 1.26 \times 1.25 = 1.57\ kg/(t\ min)$. 1.25 is ρ_{co}, kg/m^3. In accordance with the reaction $C + 0.5O_2 = CO$, 1 kg of CO is formed from 0.428 kg of C. The coke consumption $G = 0.673/0.8 = 0.841\ kg/(t\ min)$ corresponds to the intensity of CO formation equal to $1.57\ kg/(t\ min)$. When coke absorption duration is 18 min (the same as scrap heating time), the coke consumption is $0.841 \times 18 = 15.1$ kg/t, which is close to the maximum coke consumption in production practice.

Carbon monoxide introduces both physical q_{ph} and chemical q_{ch} heat into the shaft with scrap. $q_{ph} = V_{CO} \times c \times t$, $kJ/(t\ min)$, $t = 1570\ °C$ is temperature of metal during the stage of scrap melting in the furnaces with the flat bath, $c = 1.475\ kJ/(m^3\ °C)$ is heat capacity of CO at $t = 1570\ °C$. $q_{ph} = 1.26 \times 1.475 \times 1570 = 2918\ kJ/(t\ min)$ or $q_{ph} = 0.81\ kWh/(t\ min)$.

The thermal effect of the reaction $CO + 0.5O_2 = CO_2$ is 3.51 kWh/m^3 CO. $q_{ch} = 1.26 \times 3.51 = 4.42\ kWh/(t\ min)$. Heat power of the flow is $P_{CO} = q_{ph} + q_{ch} = 0.81 + 4.42 = 5.23\ kWh/(t\ min)$. The quantity of heat introduced into the shaft in 18 min is $Q = 5.23 \times 18 \times 0.9 = 84.7$ kWh/t of scrap, where 0.9 is the yield of liquid steel per 1 ton of scrap. With the heat efficiency coefficient of gases in the shaft equal 0.65, the enthalpy (heat content) of scrap E_S will amount to: $84.7 \times 0.65 = 55.0$ kWh/t. The average-mass temperature of 380 °C corresponds to this enthalpy, Table 1.2, Chap. 1.

The result of this calculation is in good agreement with the estimations of the scrap heating temperatures according to indirect data given in Sects. 2.3.1 and 2.3.2, Chap. 2. The similar calculations show that, in order to achieve the scrap heating temperature of 800 °C, the heating time in case of given operation conditions of the shaft furnace should be about 40 min.

4.3.2 Combined Use of Off-Gases and Burners; Dead-End with Regard to Energy

Insufficient heat power of off-gases flow can be compensated by the use of energy from fuel. The degree of scrap heating can be sharply increased by means of the burners. However, as is well-known, at the temperatures above 300 °C, the gases passing through the layer of scrap are intensively saturated with dioxins. For decomposition of dioxins, the off-gases leaving the shaft must be heated in a special chamber to the temperatures not lower than 1000 °C. This requires additional consumption of about 5.5 m^3/t of natural gas, which makes this system of scrap heating unreasonable in terms of energy consumption, Chap. 2, Sect. 2.3.3.

Concerning conveyor EAFs, the scrap preheating system, which uses off-gases along with burners, has been developed by Tenova Company [3]. As per this design, in front of the existing tunnel of the Consteel furnace, an additional chamber is constructed, through which the conveyor passes as well. In the roof of this chamber the burners are installed. Evacuation of gases from both the tunnel and the additional chamber is carried out via a common gas duct positioned between them. In this gas duct, after preheating a scrap located in the tunnel by off-gases, the latter mix with fuel combustion products from the burners. It is supposed that such mixing will ensure decomposition of dioxins released from the scrap being heated with the burners.

As far as energy is considered, such systems have a fundamental contradiction. For decomposition of dioxins in the common gas duct, high temperature of the gases leaving the chamber with the burners is needed in. But this high temperature is incompatible with the efficient utilization of fuel for scrap heating. In order to ensure effective decomposition of dioxins, the fuel energy in the chamber has to be used as inefficiently as the energy of the off-gases is used in the tunnel. Thus, with regard to energy, the problem of dioxins turns the combined scrap heating by both off-gases and burners into the dead-end.

The reasonable alternative is as follows. The scrap can be heated by off-gases, whereas the natural gas can be used for afterburning the dioxins in a special chamber. But this method can ensure scrap heating only to relatively low temperatures. The alternative is to use the entire temperature potential of the off-gases for decomposition of dioxins, as it is done in the conventional EAFs, and to carry out high temperature scrap heating with all its advantages by means of natural gas combustion. The latter, as the most promising method, is discussed in Sect. 4.3.3.

The requirements for ambient air protection will keep growing stricter everywhere. Therefore, using the scrap heating systems which pollute the atmosphere with dioxins is not promising even in those countries where today this is still allowable.

The system of the burners Tenova can be successfully used in the conveyor furnaces operating in the countries with cold winters. With the help of this system, ice and snow on the conveyor can be melted, and the water formed can be removed before the scarp enters the tunnel. At present, ice and snow are melted in the tunnel. Because of the low scrap temperature, the water does not have time to evaporate and enters the furnace along with the scrap. Preventing this hazard is an important task, see Sect. 2.2.2, Chap. 2.

4.3.3 Scrap Preheating by Burners Only: Replacement of Electrical Energy with Energy from Fuel

Operating shaft furnaces lag behind the best modern EAFs both in the level of productivity and in electrical energy consumption, Chap. 2, Sect. 2.4. In order to eliminate both these disadvantages it is necessary to increase a scrap preheating temperature up to at least 800 °C. The promising trend resolving the problem in complex is scrap preheating by the burners only. The mentioned trend allows avoiding additional energy consumptions for decomposing of dioxins, eliminating limitations related to insufficient heat power of off-gas flow, and opens new possibilities for developing highly productive and energy efficient steelmaking furnaces operating with scrap.

Let us determine reduction in electric energy consumption at a scrap preheating temperature of 800 °C by means of the following calculation.

4.3.3.1 Calculation

Let us examine rather the heat balance of the metal bath than the balance of the entire bath including slag. When melting down a preheated scrap its heat energy is completely absorbed by the liquid metal. Therefore, the average mass temperature of preheated scrap t_S and reduction in electrical energy consumption ΔE_{EL} obtained due to scrap preheating are connected a relationship close to unique:

$$\Delta E_{EL} = c_S \cdot t_S / \eta_{EL} \qquad (4.1)$$

c_S average calorific capacity, kWh/(kg °C)
η_{EL} the coefficient of electrical energy efficiency for heating a liquid metal

The product of $c_S \cdot t_S$ is enthalpy of scrap ΔE_{EL}, kWh/t, see Chap. 1, Table 1.2. The η_{EL} coefficient is determined by the following expression:

$$\eta_{EL} = \eta_{SEC.C} \cdot \eta_{ARC} \qquad (4.2)$$

$\eta_{SEC.C}$ the coefficient considering electrical energy losses of the secondary electrical circuit including transformer
η_{ARC} the energy efficiency coefficient of the arcs considering energy losses during the heat transfer from the arcs to the liquid metal

For the shaft EAFs operating with the flat bath and with the arcs practically continuously immersed into a foamed slag it could be assumed that $\eta_{SEC.C} = 0.93$; $\eta_{ARC} = 0.90$; $\eta_{EL} = 0.93 \cdot 0.90 = 0.84$.

Complete immersion of the arcs into the slag does not provide the η_{ARC} coefficient equal to 100 %, as is sometimes supposed. The immersed arcs transfer the heat mostly to the slag, but the slag, though it is mixed with the metal, radiates a considerable portion of the absorbed heat into water-cooled sidewall and roof panels of the furnace. The aforesaid value of the η_{ARC} and, consequently, the η_{EL} as well should be considered rather somewhat overstated than understated. This fact is confirmed by the investigation of a goodly number of modern EAFs which had shown that during the melting the η_{EL} coefficient ranges from 0.6 to 0.8 [4]. In this investigation, as opposed to the others, it is stated that the η_{EL} coefficient applies to the metal.

Given $\eta_{EL} = 0.84$, the results of the calculation according to Eq. (4.1) are demonstrated by Fig. 4.1. As the scrap preheating temperature t_S raises, the reduction in electrical energy consumption ΔE_{EL} grows at the increasing rate, curve 1. At the $t_S = 1000\ °C$, in comparison with the furnace operation without scrap preheating ($t_S = 0$), the ΔE_{EL} reaches 225 kWh/t. For example, if at the $t_S = 0$, the electrical energy consumption is 375 kWh/t, then at the $t_S = 1000\ °C$ it will decrease

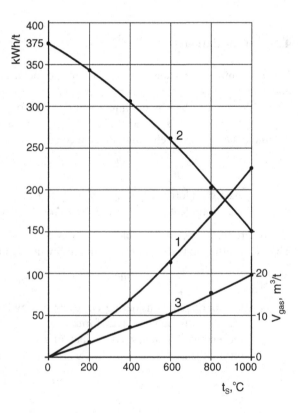

Fig. 4.1 Electrical energy consumption and gas flow rate versus average mass temperature t_S (designation in the text)

to 150 kWh/t, since the $\Delta E_{EL} = 225$ kWh/t. At the $t_S = 800$ °C, the $\Delta E_{EL} = 170$ kWh/t and the electrical energy consumption amounts to 205 kWh/t. Such electrical energy consumptions are unachievable in EAFs operating without scrap preheating.

In operating shaft furnaces, the electrical energy consumption amounts to about 300 kWh/t or somewhat higher. The value of t_S close to 400 °C corresponds to such consumption, Fig. 4.1.

The pictures of the scrap being discharged from the shaft into the window are presented by some publications. These pictures show that the temperature of the surface of scrap pieces at the discharging window can achieve about 700 or 800 °C. However, it should be kept in mind that in this zone the frontal surface of the front scrap layer is exposed to direct heat radiation from the furnace freeboard. Therefore, the mentioned temperatures are much higher than the actual average mass temperatures of the scrap located behind the window. The same applies to the lower scrap layer located on the fingers of shaft finger furnaces. If the actual average mass temperatures could reach 700 or 800 °C then the electrical energy consumption in shaft furnaces would be close rather to 200 kWh/t than to 300 kWh/t, Fig. 4.1.

Curve 3 in Fig. 4.1 shows the nature gas flow rate required to heat a scrap in EAF's shaft up to a temperature of t_S by means of burners without using the heat from off-gases. In the calculation, it was assumed that calorific capacity of gas amounted to 10.3 kWh/m^3 and the heat efficiency coefficient of gas combustion products in the shaft η_{BUR} was equal to 0.7. The thermal effect of scrap oxidation to Fe_3O_4 was also considered as follows: at the $t_S = 600$, 800 and 1000 °C that was 1.5, 2.5 and 3.5 %, respectively.

Scrap preheating by the burners only means that the use of the off-gas heat for preheating is completely abandoned. It is not easy to accept an idea of the necessity of such an abandonment since the idea as such of utilizing the heat of the off-gases for the purpose of scrap heating in the arc furnaces seems quite attractive. This idea has been dominating in steelmakers' minds for years. However, it cannot be examined in isolation from the actual possibilities of its realization in consideration of ecological problems. Persistent attempts to implement this idea in practice are being carried on for more than 30 years. Nevertheless, they did not yield the results which, in comparison with modern EAF, could justify the usage of sophisticated equipment requiring additional maintenance. The reasons why this occurs were considered above in detail. It is worth mentioning that for the reason stated above the work associated with scrap preheating for convertors has been stopped long ago, although the heat content in convertor gases is much higher as compared to that in EAF off-gases and scrap preheating would ensure a number of substantial advantages. These advantages are: shortening tap-to-tap time without an increase in bath blowing intensity, a possibility of processing heavier (i.e. cleaner) scrap, etc.

EAF's off-gas energy as well as energy of gases leaving converters and other metallurgical units has to be utilized for generation of steam. The most promising trend is to generate steam with energy parameters for the purpose of electrical power production. Lately, this trend has gained increasing attention. A good example is an integrated works in Brazil built recently by ThyssenKrupp CSA where an electrical power of 490 MW is produced by utilizing of process gases from the coke plant, blast-furnace, and BOFs. Such a quantity of electrical energy meets completely the electricity demand at the works, and around 200 MW is sold to a power utility [5]. It is worth noting that the high-temperature scrap preheating with the burners substantially increases the total heat power of off-gas flow and stabilizes this source of energy which facilitates utilizing it for steam and electricity production.

4.4 The System of High-Temperature Scrap Preheating in the Furnace Shaft by Means of Oxy-gas Recirculation Burner Devices

4.4.1 Key Features, Design and Operation of the System

The burners currently in use are unsuitable for high-temperature preheating a scrap in an EAF shaft. Air-gas burners of high power required generate too large amount of combustion products which impermissibly increase costs for their removal and purification. Oxy-gas burner flames have too high temperature, therefore, it is impossible to avoid undue oxidation, melting, and welding of scrap pieces. All this involves drop of the yield, suspension of scrap in the shaft, and fuel underfiring, Chap. 1, Sect. 1.5.2.

In order to avoid aforesaid shortcomings it is necessary to considerably reduce a temperature of combustion of natural gas with oxygen. The authors have developed the system for a deep replacement of electrical energy with fuel (REF-system), using high-temperature preheating a scrap in the shaft by means of powerful burner devices, which solves this problem due to the gas recirculation.

Oxy-gas mixture generated by the device is diluted inside it with those combustion products which have already passed through the layer of scrap, transferred heat energy to the cold scrap lowering, therefore, their own temperature. This recirculation of gases is being created by the oxygen injectors which are part of the burner's device.

The water-cooled burner device is schematically shown in Fig. 4.2. The device comprises of oxygen chamber (1) distributing oxygen to several injectors. Each injector contains oxygen nozzle (2), mixing chamber (3), and diffuser (4). In chambers (3) oxygen mixes with the combustion products which after passing through the scrap layer are sucked into the injectors via openings (5). The injectors produce a positive pressure in chamber (6) into which multiple fine jets of natural gas are fed via nozzles (7). In the chamber (6) the natural gas, oxygen, and combustion products from the shaft are completely mixed.

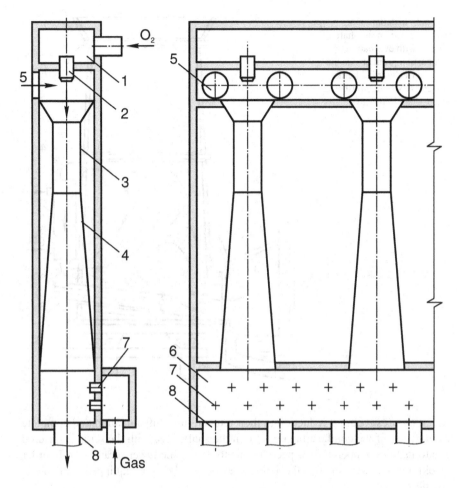

Fig. 4.2 Injector burner device (designation in the text)

The formed combustible mixture is blown under pressure via opening (8) into the furnace shaft where it is burned creating a flame with the low temperature required. The temperature of the flame decreases due to the presence in the combustible mixture of ballast in the form of the combustion products being cooled in the shaft. The correct selections of technical parameters and quantity of oxygen injectors produce required excess of pressure in the chamber (6) as well as negative pressure in openings (5).

Installation of one of burner's devices (1) on a furnace shaft is shown in Fig. 4.3. The combustible mixture is introduced in lower scrap layers inside the shaft via pipes (2) close to a pusher (3). The combustion products are sucked into the burner device via pipes (4). Thus, the combustion products before being sucked into the burner device pass through the scrap layer of a sufficient height

Fig. 4.3 Installation of burner device on the shaft. Variant with evacuation of combustion products via the furnace freeboard. *1* burner device; *3* hydraulic pusher; *2* and *4–6* are in the text

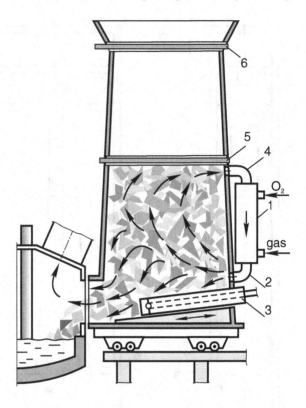

which ensures required reduction in their temperature. Inflammation and complete combustion of the combustible mixture in the shaft filled with the scrap is ensured due to its high temperature, especially due to that of the lower scrap layer. In order to eliminate backflash into the burner device the latter is equipped with flame arresters.

Along with combustion of the oxygen-gas mixture in the layer of scrap at low flame temperature, the key feature of the REF-system is stationary operating mode of the burner devices, when the heat power of the devices and the temperature of the combustion products sucked by the devices remain constant during the whole stage of scrap melting. This operating mode is optimal. It simplifies both the design of the burner devices and the process control.

The implementation of this operating mode requires that the height of the scrap layer between the level of intake of the combustible mixture through the pipes (2) into the shaft and the level of suction of combustion products through the pipes (4), Fig. 4.3, remains constant. In order to satisfy this condition, it is necessary to divide the shaft from top to bottom into two approximately equal sections. The lower section is the zone of the high temperature scrap heating, and the upper section supplies relatively cold scrap into the heating zone, as the heated scrap is pushed out from the lower section into the furnace bath.

The zones are separated with the device which allows scrap charging by the basket into the upper section of the shaft without switching off the burner devices and without gas and dust emissions into the shop atmosphere. A mobile hopper with opening bottom serves as such a device in the scrap charging system EPC, Chap. 2, Sect. 2.3.2. In the ESS system of the CVS company, Turkey, the upper section of the shaft is separated from the lower section by a gate (5). Together with the second gate (6) closing the shaft from the top, this system transforms the upper section of the shaft into a lock chamber. The scrap is charged into the shaft when the upper gate (6) is open and the lower one is closed. Then the upper gate is closed, and the lower one is opened. As soon as all the scrap from the top section of the shaft moves into the bottom section, the lower gate is closed, and scrap charging into the top section of the shaft is repeated.

It should be noted that the described above REF-system has one additional important advantage over the conveyor scrap charging. Overheating of the lower layer of scrap, where the combustible mixture is blown in, does not impede reliable operation of the pusher. This is confirmed by the experience obtained from operating the shaft COSS-type furnace in China, in which the scrap temperature reached 1000 °C due to the fact that the content of hot metal in metal-charge was 40 %, Chap. 2, Sect. 2.3.2.

Besides that, in the system with the pusher, the scrap can be charged into the bath by multiple small portions rather than by three quite large portions as in the finger shaft furnace Quantum, Chap. 2, Sect. 2.3.1. In case of small portions and thin scrap layer, the average temperature is considerably closer to the maximum temperature than in the portions with thicker layer. Consequently, higher average-mass scrap temperatures can be achieved in the furnaces with the pushers than in the finger shaft furnaces using any heating method. The REF scrap preheating system is more detailed examined under operational conditions of fuel arc furnace FAF, Chap. 6. It should be noted that this system can be used in the Quantum furnaces as well.

4.4.2 Gas Evacuation and Environment

In the REF-system, the physical and chemical heat of the process gases (CO) evolving from the furnace bath is used for decomposing of dioxins which saturate combustion products in the furnace shaft. For this purpose, the combustion products leaving the shaft are mixed up with the high-temperature process gases either in the furnace freeboard or in a special post-combustion chamber. The first option [1] requires a sharp increase of size and current-capacity of a roof elbow and, for this reason, is not always acceptable. The second option is more universal and does not limit the power of burner devices, as opposed to the first option.

On their way from the roof elbow to the combustion chamber, the process gases are not diluted by atmospheric air, which allows to most fully preserve their energy potential. Only the minimum quantity of air required for the full CO post-combustion is infiltrated into the chamber, which ensures the maximum combustion temperature. In the chamber, the high-temperature process gases are mixed up with the combustion products entering from the shaft, which results in decomposition of dioxins. The atomized water is injected into the gas duct connecting the chamber with the bag filters, which sharply reduces the temperature of the off-gases to the level excluding the possibility of the dioxins recombination.

The system under consideration can ensure environmental protection from the emissions of both CO and dioxins complying with the strictest modern standards. However, it imposes high requirements on the combustion chamber design. It is necessary to provide complete mixing of the flows of the gases entering the chamber, high turbulence level of these flows, and sufficiently prolonged time for which the mixture of gases stays in the chamber at temperatures close to 1000 °C. A case is known when successful solution of this complex problem required two consecutively positioned post-combustion chambers [3].

References

1. Toulouevski YN, Zinurov IY (2013) Innovation in electric arc furnaces. Scientific basis for selection, 2nd edn. Springer, Berlin
2. Abel M, Dorndorf M, Hein M et al (2011) Highly productive electric steelmaking at extra low conversion costs. MPT Int 3:92–96
3. Memoli F, Guzzon M, Giavani C et al (2011) The evolution of preheating and the importance of hot heel in supersized Consteel systems. In: AISTech conference, proceedings, vol 1. Indianapolis, USA, pp 823–832, May 2011
4. Pfeifer H, Kirschen M, Simoes JP (2005) Thermodynamic analysis of EAF electrical energy demand. In: 8th European electric steelmaking conference, Birmingham, May 2005
5. Wurth P (2010) Inauguration of the Thyssen Krupp CSA integrated iron and steel works in Brazil. MPT Int 4:30–32

Chapter 5
Increasing Scrap Melting Rate by Means of Bath Blowing

Abstract Various ways and devices for oxygen bath blowing are analyzed from the point of view of their influence on a scrap melting rate in liquid metal. For the purpose of the highest increase in the scrap melting rate, the bath blowing system with movable roof oxygen tuyeres developed by the authors is offered. Submerging of tuyeres to slag-metal interface as well as direction of jets to the scrap charging zone are key features of the system. Such directional submerged blowing can provide considerable increase in the intensity of metal stirring at the scrap charging zone and sharp shortening of scrap melting time as well. Basic principles of calculation and design of oxygen tuyeres having a durability required are examined. The offered roof tuyeres comprise of tips with most effective jet cooling which provides high tuyere durability when submerging not only into slag but also that into metal. Basic designed tuyere parameters are given. The tuyeres are automatically maintained at the slag-metal interface by using reliable sensors monitoring positions of tuyere tips relatively to this interface.

Keywords Devices for oxygen bath blowing · Jet modules · Converter type tuyeres · Tuyeres cooled by atomized water · Roof tuyeres · Submerging into the melt to slag-metal interface · Durability of tuyeres · Controlling position of tuyeres · Jet cooling of tuyere tips · Increase in scrap melting rate · Methods of calculation of tuyeres

5.1 Purpose of Blowing

On the modern EAFs with scrap charging from the top, bath blowing is mainly aimed at solving a major task which basically is energy related, i.e. ensuring the fastest possible charge melting and metal heating to the tapping temperature. For this purpose, the blowing intensity is increased, which increases the bath stirring rate, as well as its melting, heating, and decarbonizing. Evolving from the bath carbon monoxide burns up in the freeboard and heats the scrap, which contributes to further acceleration of its melting.

To increase the amount of CO and reduce iron oxides, the carbon powder is blown into the bath simultaneously with oxygen blowing. This allows to even further increase the intensity of oxygen blowing without increasing the metal oxidation and decreasing the yield. The intensive oxygen blowing also helps equalizing temperatures over the depth of the metal bath. The temperature of the lower layers of metal increases, which accelerates the melting of large scrap pieces lying on the bottom of the furnace.

The shaft furnaces with flat bath and scrap charging by pushers add some new essential aspects to the purposes of bath blowing. There is no scrap in the freeboard of these furnaces. Therefore, post-combustion of CO above the bath becomes not important. The main purpose of blowing in this case is to shorten the duration of scrap melting in the restricted zone of its charging due to increasing of stirring intensity and of velocity of metal motion in this zone. Such approach to this problem is determined by two factors. First of all, insufficient scrap melting rate in the charging zone is the bottleneck of this process limiting its productivity. Secondly, as it was shown earlier, scrap melting rate in liquid metal strongly depends on the speed of metal flows around the scrap pieces, Chap. 3, Sect. 3.5.1, Eq. (3.10).

The only practically available way of increasing speed of metal flows and of stirring intensity in the charging zone is to use the blowing devices specifically designed for solving of this problem; and the selection of these devices should most closely correspond to the specific purpose.

5.2 Selection of Methods and Means of Blowing

5.2.1 Jet Modules

At present, jet modules for oxygen bath blowing in EAFs became widespread over the world. These are the multifunctional devices. In various combinations, they fulfill functions of burners used for scrap heating, oxygen lances, injectors used for carbon/lime injecting into the melt. All structural elements of the modules are usually placed in water-cooled boxes protecting these elements from high temperatures as well as from damage during scrap charging from above. The boxes are inserted into the furnace through the apertures in the sidewall panels, which decreases the distances from the burners and from injectors to the bath surface. There is a wide variety of design versions of the jet modules. The advent and development of this trend is associated with the PTI Company (USA) and with the name of V. Shver. Let us examine the typical design of the module.

The oxy-gas burner (2) with water-cooled combustion chamber (3) and pipe (4) for the injection of carbon powder are located in the water-cooled copper box (1), Fig. 5.1. The burner (2) has two operating modes. In the first mode, it is used as

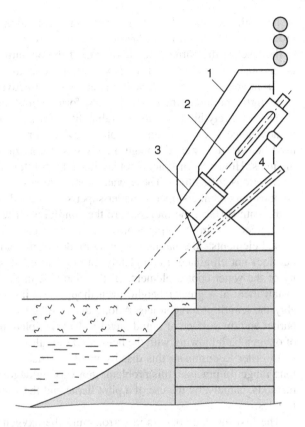

Fig. 5.1 Jet module (designations are given in the text)

a burner for heating of scrap and operates at its maximum power of 3.5–4.0 MW. The gas mixes with oxygen and partially burns within the combustion chamber (3). At the exit from this chamber, the high-temperature flame is formed, which heats and caves-in intensively the scrap in front of the burner. The combustion chamber protects, to a considerable extent, the burner nozzles from the clogging by splashes of metal and slag.

In the second operating mode, the burner is mainly used as a device for blowing of the bath. The gas flow rate considerably decreases, and the oxygen flow rate sharply increases. In this case a long-range supersonic oxygen jet is formed. In this mode, the function of the burner alters. It is reduced to the maintenance of the low-power pilot flame. This flame surrounds the oxygen jet increasing its long range.

The burner is controlled by a computer which switches its operating modes in accordance with the preset program. Immediately after scrap charging, the first mode is switched on. In several minutes, it is switched to the second mode. The highly heated scrap can be cut by oxygen considerably easier than cold scrap. Therefore, the preliminary operation of the burner in the first mode greatly facilitates penetration of the supersonic oxygen jet through the layer of scrap to the hot heel on the bottom. This ensures the early initiation of blowing of the liquid metal with oxygen, which is the necessary condition for achievement of high productivity of the furnaces. While the upper layers of scrap continue to descend to the level of the burner, the alternation of the operating modes is carried on and is repeated after charging of the next portion of scrap. In EAFs with scrap charging from above, this considerably increases the effectiveness of the use of oxygen in the initial period of the heat before the formation of the flat bath.

The relatively high operational reliability of the jet modules is their fundamental advantage in comparison with other devices for the bath blowing with oxygen used before. Just this advantage mainly ensures a wide spread occurrence of jet modules. However, in practice obtaining a high reliability of boxes is associated with serious difficulties. These water-cooled boxes and the fronts of the burner combustion chambers operate under super severe conditions. Moreover, the closer to the bath surface, the more severe the conditions. The blow-back of the oxygen jets reflected from the scrap lumps are the main cause of damage of these water-cooled elements. Alternating operating modes of the burner reduces this problem, but does not eliminate it completely. In order to increase the insufficient durability of the water-cooled elements of the modules, in many cases, it is preferred to install them at a greater height, even though this installation increases considerably the length of oxygen jets. A long distance between the oxygen nozzle of the burner and the surface of liquid metal is the principled disadvantage of the method of oxygen bath blowing with the help of jet modules.

In order to eliminate this disadvantage it is necessary to increase oxygen jets long range. In practice, this problem, in a varying degree, is solved by two known methods, namely, by the use of a pilot flame and the so-called coherent supersonic nozzles.

The first method consists in surrounding the oxygen jet with the annular pilot flame of the burner. The gases in the flame have low density due to the high temperature. In this case, the known effect of increase in the range of the jet flowing into the less dense ambient medium is used.

However, let us also take into consideration a fact that, even without the surrounding pilot flame, the oxygen jets propagate in the EAF freeboard filled with gases with the temperature in the order of 1600 °C and higher. These gases contain a large amount of CO. Taking into account post-combustion of CO in the O_2 jets, the density of the gases is not much different from the gas density in the pilot flame. Therefore, the additional increase in the long range of oxygen jets resulting from surrounding the jets with the pilot flame cannot be considerable.

The general use of the pilot flame in the modules, which requires significant additional consumption of natural gas, is required not so much for increasing the range of oxygen jets, as for its ability to facilitate penetration of oxygen jet to the melt through the layer of scrap. At the same time, the probability of reflection of oxygen jets from the scrap lumps to the water-cooled elements of the modules decreases.

However, such an effective function of the pilot flame is excluded in the furnaces with flat bath since scrap is absent in their freeboard. In respect of increasing the long range of oxygen jets, there are no reliable data obtained under industrial conditions, which confirm an improvement in effectiveness of oxygen bath blowing, due to that increasing. At the conveyor furnace of 170 t capacity, Asha, Russia, the abandonment to use the pilot flame, when blowing the bath with jet modules, allowed reducing natural gas flow rate and had no effect on the furnace operation performances.

The second method of increasing the jets long range consists in configuring the optimum profile of the supersonic nozzle, which minimizes the turbulence of the flow. By analogy with the lasers, these nozzles and the supersonic jets formed by them are called coherent. Due to the quite low turbulence, the coherent jets involve into their motion considerably smaller mass of the ambient gas, expand slower, and maintain initial velocity at the considerably greater distance from the nozzle, in comparison with the jets flowing from the simple de Laval nozzles.

In the tests conducted in the laboratories on the testing ground, the achieved increase in the length of the initial region of the coherent jets was by approximately 5 times more in comparison with that of the jets flowing from the simple de Laval nozzles.[1] Such results of the stand tests can make a false impression regarding the possibility of quite significant increase in the distances from the jet modules to the bath surface. However, it is necessary to understand that the conditions of the coherent nozzles operation in the EAF differ considerably from the laboratory conditions. Even the simple de Laval nozzles are quite sensitive to the most insignificant deviations of the oxygen parameters from the design parameters. To an even greater degree, this relates to the coherent nozzles. It is noted that small deviations from the design operating conditions eliminate the advantages of these nozzles, whereas under the production conditions, such deviations occur constantly. Furthermore, even quite small deposits of droplets of metal and slag inside the nozzles, which cannot be avoided in practice, cause strong turbulization of the flow disrupting its coherence. It is also necessary to note that there are no any direct or indirect data obtained under the actual conditions in the furnaces, which could confirm the results of the stand tests. All the aforesaid leads to the conclusion that the known methods of increasing the long range of oxygen jets cannot compensate a large distance of the jet modules from the surface of the metal bath. It is impossible to sharply increase the intensity of bath stirring in the scrap charging zone of EAFs with flat bath by means of jet modules. For this purpose some different more effective devices have to be used.

5.2.2 Converter Type Bottom Tuyeres

As is known the most intensive stirring can be ensured by oxygen tuyeres installed in the bottom of a furnace. Similar devices are used in oxygen converters with bottom blowing. Taking into consideration the experience of bottom blowing

---

[1] Let us recall that initial region is the region of the jet where its axial velocity does not decrease and remains equal to the initial velocity.

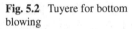

Fig. 5.2 Tuyere for bottom blowing

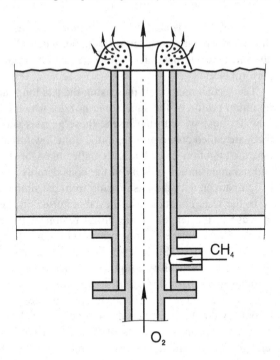

accumulated in the oxygen converters, Klöcker Company, Germany, in the nineties has developed the technology of EAF's steel melting with blowing of the bath from below, which has been called the K-ES. In order to implement this technology, several blowing tuyeres of the pipe-in-pipe type are installed in the furnace bottom. The design of these tuyeres is analogous to that of the tuyeres for the bottom blowing in the oxygen converters, Fig. 5.2. Oxygen is delivered through the inner pipe, while natural gas or other hydrocarbon fuel is delivered through the annular gap. The oxygen jet entering the liquid metal is enveloped by gas, which prevents direct contact of oxygen with the bottom lining. Natural gas entering the high-temperature zone undergoes carbonization and forms at the tuyere exit a porous carbonaceous build-up ("a mushroom") which protects the refractory lining against intensive wear.

In accordance with the K-ES technology, simultaneously with the bottom oxygen blowing, powder carbon is injected into the bath, in some way or other. A portion of carbon in the form of coke is charged into the furnace together with the scrap. Carbon monoxide (CO) evolving from the bath in large quantities is post-combusted above the bath with the help of the oxygen tuyeres installed in the side wall panels along the entire perimeter of the furnace. The wall oxy-gas burners are used for the scrap preheating. In the liquid bath stage, more oxygen is added for the post-combustion of CO. The total oxygen flow rate for blowing of the bath and post-combustion reaches 55 m^3/t, and carbon consumption is 27 kg/t. The preponderant amount of oxygen is used for post-combustion of CO, and the share of carbon injected into the bath is equal to about 50 % of its total consumption.

The use of this technology and of some of its elements in a number of plants in Italy and other countries has ensured a significant decrease in tap-to-tap time, as well as in electrical energy consumption. As in the oxygen converters, the bottom blowing has sharply improved the bath stirring which has decreased the content of oxygen in the metal and of FeO in the slag, and has increased the yield. The intensity of the bottom oxygen blowing in the K-ES process has been limited by relatively small depth of the furnace bath. Therefore, the obtained results have not been substantially different from those achieved by other methods without serious difficulties related to installation of bottom tuyeres in bottom lining.

Bottom tuyeres weaken the bottom and increase a risk of metal escape. Under actual operational conditions, monitoring a bottom wear degree by means of existing facilities is not reliable enough. Hot repairs of the bottom during the furnace operation take a lot of time. Therefore, at some plants, for example, at Ferriere Nord SpA, Italy, the bottom is replaced in the specified number of heats regardless of a visible wear degree of it. For this purpose, permanently, there is a reserve complete set (stand-by) of the bottom for immediate installation on the furnace. Due to the proper maintenance organization replacement of bottoms takes a little time.

With the use of bottom oxygen blowing, the bottom wear rate is markedly higher than that in conventional EAFs. Therefore, in the furnaces with bottom tuyeres, in the majority of cases, instead of oxygen for bath blowing argon, nitrogen, or their mixtures are used depending on the grades of steel. Although such blowing is carried out with low intensity of the order of 0.1 m^3/t min, it appears to be quite effective, because it considerably improves the bath stirring. In spite of this, bottom blowing became quite limited spread only because of operational difficulties discussed above. This is the reason why it cannot be considered as an optimum alternative for furnaces with the flat bath as well, although positioning of bottom tuyeres close to a zone of scrap charging could sharply increase the rate of scrap melting.

5.2.3 Tuyeres Cooled by Evaporation of Atomized Water

It is necessary to minimize the distance between tuyere nozzles and a liquid metal surface so that oxygen jets could intensively stir the metal bath. Maximum effect is ensured by the immersion of oxygen tuyeres into the melt down to slag-metal interface. This is confirmed by both the results of simulation and long-term international experience of blowing a bath of open-hearth furnaces [1]. In open-hearth furnaces, the submerged blowing has overall replaced blowing with tuyeres positioned above the bath as soon as roof tuyeres which possess high enough durability were developed. It should be emphasized once again, that hydro-dynamical processes in the bath of hearth furnaces such as open-hearth and electric arc furnaces are completely similar. Therefore, the open-hearth experience of the bath blowing is of a great interest for modern EAFs as well.

In the second-half of the nineties, understanding the fundamental advantages of the submerged blowing resulted in the development by the KT-Köster Company, Germany, of the blowing tuyeres of a new type, the so-called KT-tuyeres. These tuyeres are installed in the lining of the banks of bottom of the furnace. According to the initial concept, the oxygen KT-tuyeres had to be installed slightly lower than the slag surface, and the KT-tuyere for the carbon injection even lower, i.e. near the slag-metal interface. The water for KT-tuyere cooling is used in the mixture with the compressed air. The water is atomized in the heat-stressed frontal part of the tuyere near the head. The small drops of water evaporate on the extended surface of the head. There is no water in the tuyere head itself. In case of the tuyere burn-back, only a small amount of highly atomized water can get into the liquid metal which is no danger.

In the KT system, the heat is mainly removed rather due to evaporation of water than due to heating of it as in the usual water-cooled devices. It is known that the quantity of heat required to heat water from 0 °C to the boiling point of 100 °C is approximately 5 times less in comparison with the quantity of heat required for its total evaporation at this temperature. This allows reducing water consumption for cooling of KT-tuyeres multi-fold as compared to that for cooling of jet modules. The mixture of air, residual portion of atomized water, and water vapor is sucked off from the tuyere by the vacuum pump. As a result, there is no excess pressure in the tuyere. After exiting the tuyere, the air separates from the water and exits into the atmosphere, the vapor condenses, and water is returned into the cooling system of the furnace.

Installation of stationary water-cooled tuyeres and other water-cooled elements in the lining of EAFs' bottom banks even higher than the level of the sill of slag doors is with good reason considered as highly dangerous. The accumulated practical experience shows that, in this case, visible burn-backs of the tuyeres do not pose a special hazard. The visible burn-backs are rapidly detected due to a number of signs indicating the burn-backs, which allows to promptly stop the water leakage by switching off the burnt tuyere. Small hidden leakages, which are difficult to detect on time, pose a real danger. Such leaks damage the internal, hidden from observation lining layers resulting in severe accidents with explosions and metal escape. Therefore, the installation of water-cooled tuyeres in the bottom banks lining must be used only with the application of highly reliable technical equipment, which either guarantees practically inertialess detection of any smallest hidden water leakages or eliminates any possibility of such leakages. The KT-system completely satisfies this requirement especially due to a negative pressure inside the tuyere. A complete safety of installation in the bottom banks lining can be ensured by the tuyeres cooled by high-pressure water circulating in a short closed circuit as well [2]. However, such very promising systems, unlike the KT-tuyeres, have not yet found a practical application in EAFs.

The industrial trials of the KT-tuyeres have shown that the cooling system utilized in them does not allow the contact of the tuyere head with the liquid metal. Therefore, later on, not only oxygen KT-tuyeres have been installed only into the slag close to its surface at the sufficiently great distance from the slag-metal interface, but carbon KT-tuyeres as well. As the slag foaming takes place, the heads of the KT-tuyeres are submerged into the slag to a significant depth. It is impossible to install the blowing devices of the jet module type at that height.

Compressed air is used for injecting of carbon powder into the slag. Natural gas, which is introduced through the annular gap surrounding the oxygen jet, is delivered to the oxygen KT-tuyeres. The flow of gas protects the refractory from the contact with oxygen and, thus, prevents rapid wear of the lining. Compressed air and oxygen provide protection from the slag flowing in the KT-tuyeres. Similar to jet modules, oxygen KT-tuyeres are used for blowing of the bath during the liquid bath stage, and used as burners for scrap heating at the beginning of the heat. Oxygen and carbon KT-tuyeres are installed in pairs, next to each other. These pairs are dispersed along the entire perimeter of the bath. The oxygen, carbon, natural gas and compressed air feeding to the tuyeres is fully automated.

KT-tuyeres have found a certain use. The obtained results confirm a sharp increase in the effectiveness of blowing when the tuyeres are submerged into the slag. In this respect, the most demonstrative are the data obtained from 100-t DC EAF operating on a charge material containing from 80 to 90 % of metalized pellets [3]. The pellets are charged into the bath by a conveyor through the opening in the furnace roof.

In 2003, manipulators with consumable pipes were replaced with the KT system. Four oxygen and two carbon KT-tuyeres were installed. Before the installation of these tuyeres, it took 60 min to charge 65 t of pellets. It was impossible to charge the pellets faster, because this led to the formation of large "icebergs" from the unmolten pellets floating on the bath surface. When working with the KT system, 74 tons of pellets are charged in 33 min. In this case, the furnace productivity with regard to the pellets melting process increased more than doubled.

These results could be obtained only in case of radical increase in the intensity of bath stirring by oxygen jets due to submerging of the heads of the KT-tuyeres into the slag. The same can be said about the results obtained from the 100-t EAF of another plant operating on scrap. Installation of the KT-tuyeres has led to a sharp acceleration of melting of the large lumps of scrap submerged in the liquid metal. This has allowed to charge into the furnace considerably larger lumps of scrap and, thus, to decrease the cost of scrap preparation for melting without decreasing the productivity.

These results could be obtained only in case of radical increase in the intensity of bath stirring by oxygen jets due to submerging of the heads of the KT-tuyeres into the slag. The same can be said about the results obtained from the 100-t EAF of another plant operating on scrap. Installation of the KT-tuyeres has led to a sharp acceleration of melting of the large lumps of scrap submerged in the liquid metal. This has allowed to charge into the furnace considerably larger lumps of scrap and, thus, to decrease the cost of scrap preparation for melting without decreasing the productivity [4].

The rate of pellets melting is determined exclusively by intensity of heat transfer to them from liquid metal. This can be explained by the fact that due to a small size of pellets their internal thermal resistance to the transfer of heat is very low as well. The same relates to furnace scrap owing to a small thickness and high thermal conductivity of its lumps. Therefore, it can be supposed that installation of the KT-tuyeres in the scrap charging zone of furnaces with the flat bath would result in a sharp increase in the rate of scrap melting when the KT-tuyeres are constantly submerged into the slag. The necessity of submerged blowing when melting a scrap in the liquid metal is convincingly confirmed by the results obtained from KT-tuyeres applications.

In addition to advantages resulted from the stirring intensification, the submerged blowing of the bath by the KT-tuyeres has increased the durability of the central refractory part of the furnace roof by 2–3 times, consequently, the consumption of refractory materials and repair costs have been reduced. This is explained by the fact that the durability of the roof refractory is directly related to intensity of bath splashing which is decreased by many times when submerging the tuyeres into the slag. Such a relationship is persuasively confirmed by the results of the studies of splashing conducted by the methods of physical modeling [1].

These studies were carried out on a model of the triple-nozzle vertical tuyere in the full compliance with the requirements of the theory of similarity. Liquid steel was simulated by a solution of iodide potassium, and slag was simulated by a solution of vacuum pump oil in kerosene. The ratio between the densities of both solutions complied with the similarity conditions. A nozzles downward angle to the vertical was 30°. As the tuyere approaches to the bath surface and, after its submerging into the slag, to the slag-metal interface, the following changes in the intensity of splashing are being observed, Fig. 5.3. With a large distance between the nozzles and the slag surface, the impact of the oxygen jets on the bath is minor and the splashing intensity is negligible. At a certain small distance from the slag surface, the splashing intensity increases rapidly and reaches the maximum. At that moment the slag produces more splashes than the metal does. At even closer distance between the nozzles and the bath surface, the splashing intensity reduces sharply, and with the nozzles submerged into the slag, it reduces practically to zero. As the nozzles, after further submerging, get even closer to the slag-metal interface the splashing intensity increases again. The second maximum of splashing intensity appears, but its magnitude is so minor that it can be neglected. The splashing stops altogether as soon as the tuyere submerges into the metal.

Despite the full compliance with the similarity conditions, splashing of the bath on a cold model noticeably differs from that under the actual conditions of oxygen blowing. When submerging tuyeres into the slag, splashing does not stop completely although diminishes many-fold. This is confirmed by high-speed video recording and other methods of studies carried out on hearth-open furnaces [1]. Unfortunately, in EAFs such studies have not been carried out.

However, in case of the KT-tuyeres installed in the slag, the evaporative cooling does not provide their sufficiently high durability. Although the design of KT-tuyeres allows for the possibility of replacing the worn part of the head without dismantling of the entire tuyere, their maintenance costs, in comparison with the jet modules proved to be quite significant. This has largely limited their wide use.

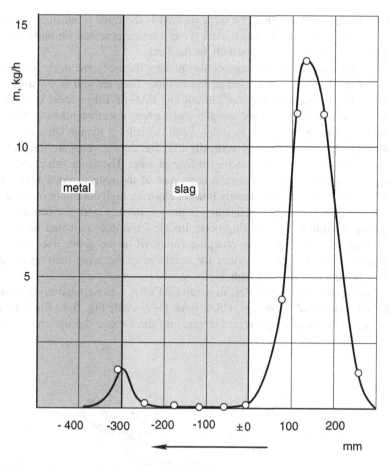

Fig. 5.3 Dependence of splashing intensity, m, kg/h, on position of tuyere nozzles relatively to the slag surface (results of physical simulation [1]). The *arrow* indicates direction of motion of the tuyere

5.2.4 Mobile Water Cooled Tuyeres

These tuyeres are inserted into the freeboard with the help of the manipulators through the slag door or through the openings in the sidewalls and the roofs of the furnaces. During the first stages of implementation of these devices in the Soviet Union and in a number of other countries, vertical roof tuyeres were used. This method was taken from the open-hearth furnace operation. However, due to more severe operational conditions in the EAFs in comparison with the open-hearth furnaces, the durability of the roof tuyeres of the open-hearth type proved to be insufficient. The tuyeres have frequently burnt through. Because of this and other operational difficulties caused by positioning of the tuyeres on the furnace roof, the application of this blowing method has soon ceased. Nevertheless, as it is shown in Sect. 5.4, this bath blowing method is the most promising, providing that its deficiencies are eliminated, since it has the best potential for increasing the scrap melting rate in the furnaces with the flat bath.

The door-mounted mobile water-cooled tuyeres have become more widespread and have been used over prolonged period of time. They are still in use on a number of the EAFs. For example, on 420-ton DC EAF of Tokyo Steel company in Japan, the twin door tuyere for oxygen and carbon injection into the bath was installed in addition to four jet modules. With the help of manipulator, the tuyere can rotate both horizontally and vertically [5]. The door tuyeres are inserted into the freeboard at a small angle to the surface of bath. Therefore, when the tip of the tuyere submerges into the slag, a large part of the long external tuyere pipe submerges as well. This significantly hinders ensuring high durability of these tuyeres. In addition, their other deficiency is incompatibility with the furnace operational mode with the closed slag door. Besides, the door-mounted tuyeres are located quite far from the scrap charging zone. All of the above does not allow considering these tuyeres as devices for accelerating the scrap melting in liquid metal in the furnaces with flat bath.

Before creating of jet modules, in many EAFs the water-cooled tuyeres developed by Berry Metal Company, USA, have been used, Fig. 5.4. These tuyeres characterized by large length were inserted into the furnace through the openings

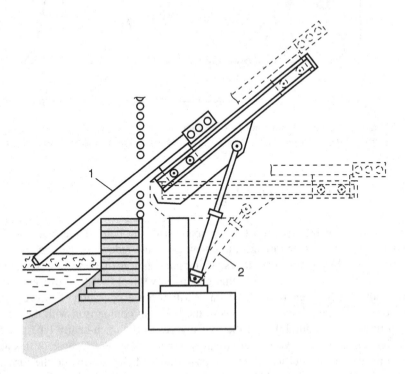

Fig. 5.4 Mobile water-cooled tuyere, *1* tuyere, *2* hydro-cylinder

in the side wall panels at an angle to the bath surface of about 40°. At the end of the blowing, the tuyeres were removed from the freeboard and lowered back into the horizontal position. Usually, two tuyeres were installed in the furnace. The tips of the tuyeres could have two oxygen supersonic nozzles and one channel for the carbon injection positioned parallel to the axis of the tuyere. With respect to this axis, the oxygen nozzles were positioned so that the oxygen jets attacked the bath at an angle of about 30° to the vertical.

The tuyeres have been developed for the submerged blowing whose advantages were already well known from the operational experience gained from the open-hearth furnaces. However, the tuyeres in the EAF could not withstand such a mode of operation because of their design drawbacks. Submerging of the tuyeres into the slag caused so frequent burn-backs of the tuyeres that, according to the report of the Nucor Yamato Steel Company, USA, the costs on their tips replacement for two furnaces amounted to $400,000 a year [6].

Because of low durability of tuyeres, the operators avoided submerging the tuyeres into the melt and kept the tuyeres mostly above the slag during the blowing. Such a mode of use, to a considerable extent, has eliminated the potential advantages of the mobile water-cooled tuyeres. Eventually, in the majority of cases, this has resulted in an abandonment of their application and general use of jet modules. Thus, the unresolved problem of durability of the mobile water-cooled tuyeres has led to switching from potentially more effective method of the submerged blowing to known to be less effective, but more simple and reliable in operation method of bath blowing from the top.

5.3 Principles of Design and Calculation of Highly Durable Water Cooled Tuyeres

5.3.1 Heat Operation of Tuyeres, Fundamental Dependences, Heat Flows, and Temperatures

Being submerged into a liquid bath, a tip of the tuyere and a part of its side surface are affected by the heat flows of very high density. If the necessary conditions of heat transfer are not observed, these heat flows, when passing through the tuyere surface to water, can cause impermissible overheating of the submerged part of the tuyere, its rapid wear and burnout.

Under stationary conditions, when the total amount of heat obtained by the tuyere is consumed for heating of water, the thermal operation of the submerged part of the tuyere is determined by the following dependences: the heat balance Eq. (5.1) and the heat transfer Eqs. (5.2)–(5.4) and Fig. 5.5.

$$Q = q_m \cdot F_1 = q_w \cdot F_2 \tag{5.1}$$

$$q_m = \alpha_m(t_m - t_1) \tag{5.2}$$

$$q_\lambda = \lambda(t_1 - t_2) / \delta \qquad\qquad (5.3)$$

$$q_w = \alpha_w(t_2 - t_w) \qquad\qquad (5.4)$$

$Q; q_m$ heat flow, W, from melt to tuyere surface, and density of heat flow, W/m^2

F_1 surface area of submerged part of the tuyere, m^2

q_w density of heat flow from wall to water, W/m^2

F_2 area of the wall contacting water, m^2

α_m coefficient of heat transfer from melt to tuyere surface, W/(m^2 °C)

t_m temperature of melt, °C

t_1, t_2, t_w temperatures, °C, of tuyere surface, of surface contacting with water, and of water, respectively

q_λ density of heat flow in the wall, W/m^2

λ coefficient of thermal conductivity of the wall, W/(m °C)

δ wall thickness, m

α_w coefficient of heat transfer from wall to water, W/(m^2 °C)

The Eq. (5.1) shows that the process under consideration is a stationary process in which the amount of heat accumulated by the tuyere does not vary. Under the stationary conditions, the entire heat flow obtained by the tuyere surface from the melt (amount of heat per time unit) is transferred to water. It follows from the Eq. (5.2) that the heat flow from the melt per unit of the tuyere surface (density of the heat flow) is determined by the external factors and does not depend on the parameters of the tuyere itself. The Eq. (5.3) explains the heat transfer through the tuyere wall when the wall thickness δ is small and its curvature can be neglected. In actual practice, the tuyeres for bath blowing always satisfy this condition. Finally, the Eq. (5.4) determines the tuyeres parameters which affect the intensity of heat transfer to water.

Typically, the heat flow density q_m is distributed over the tuyere surface non-uniformly. Usually, the maximum values of q_m are observed near the oxygen nozzles and in the places exposed to blow-backs of oxygen jets reflected from the unmolten scrap pieces. When the tuyeres submerging into the melt come into contact with the scrap, it often causes their burnout.

Under the actual conditions of the bath blowing by the submerged tuyeres, it does not seem possible to determine the values of q_m analytically with acceptable accuracy. Therefore, the experimental data should be used in calculations. Unfortunately, measurements of q_m under operating conditions of submerged blowing were conducted only in the open-hearth furnaces. The most reliable data are given in the publication [1].

In this work, the density of the heat flow on the submerged part of the tuyere was determined as the difference between the total heat flow on the entire tuyere and the flow on its side surface outside of the melt. The total flow was found by measuring the flow rate and temperature of water, and the flow on a side surface was determined by measuring the elongation of the external pipe of the tuyere with due correction for the depth of its submerging. The obtained values of

q_m were 1300–1500 W/m^2 in the slag and 4000–5000 W/m^2 in the metal. Triple or even quadruple increase of q_m in the metal is mainly explained by the respective increase of α. In the open-hearth furnaces and in the modern arc furnaces with flat bath, the temperatures of the bath during scrap meting stage are approximately the same. Therefore, the value of q_m given above can be used for calculating the parameters of the tuyeres for submerged blowing in the furnaces with scrap melting in liquid metal.

Let us review the temperature characteristics of heat operation of the tuyeres. The most reliable indicator of the potential durability of the tuyere is the temperature of its external surface t_1, Fig. 5.5. Rather prolonged period of service is possible if the values of t_1 are low. And the rapid wear of tuyere usually occurs in case of high temperature of its surface. Let us analyse the factors affecting this parameter.

Let us convert the Eq. (5.3) to the equation of the form

$$t_1 = t_2 + q_\lambda \cdot \delta / \lambda \tag{5.5}$$

Fig. 5.5 Tuyere for coke injection. *1* Bath surface, F$_2$ and F$_2$—areas of surfaces shown by *heavy lines*

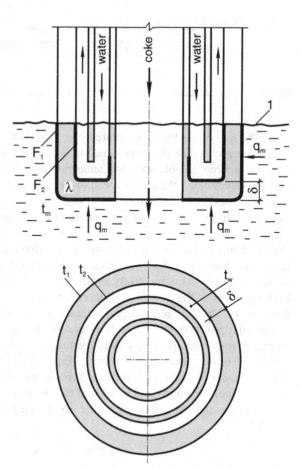

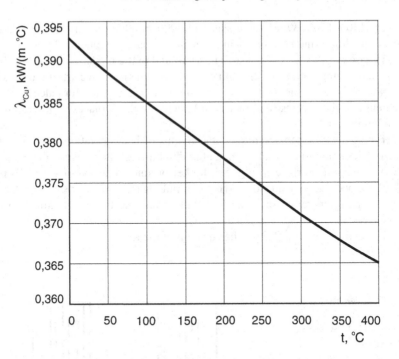

Fig. 5.6 Thermal conductivity λ_{Cu} of copper versus temperature t

The value $q_\lambda \cdot \delta/\lambda$ is the temperature differential in the tuyere wall $\Delta t = t_1 - t_2$. This value grows directly proportional to q_λ and to the thermal resistance of the wall δ/λ. In order to obtain the low temperature of the tuyere surface t_1 in case of the high values of q_λ, the external wall of the submerging part of the tuyere should be made of copper of the highest purity and density which has the maximum thermal conductivity. The dependence between the coefficient of thermal conductivity of such copper and the temperature is given in Fig. 5.6. The wall thickness δ should be minimum allowed by condition of satisfying the strength requirements. Increasing the wall thickness as a reserve is unreasonable, because this results in the increase of t_1 and, respectively, of the wear rate, which surpasses "the reserve". High durability of the tuyeres is achieved when the surface temperatures are lower than 200 °C. The wear rate sharply increases at $t_1 > 300$ °C.

Ensuring low enough temperature level of the inner surface of the wall t_2 is especially important for durability of the tuyeres. This temperature is not just one of the addends determining the value t_1, Eq. (5.5). It also affects the mechanism of heat transfer from the wall to water. Variations in this process can lead to sharp acceleration of wear and rapid burnout of the element.

The temperature t_2 is defined by the expression (5.6) which follows from the Eq. (5.4):

$$t_2 = t_w + q_w/\alpha_w \qquad (5.6)$$

The Eq. (5.6) shows that the temperature t_2 depends not only on the water temperature t_w and the coefficient of convective heat-transfer α_w, but on the density of the heat flow from the inner surface of the wall to water q_w as well. Despite the fact that, in case of stationary mode, the same heat flow Q invariably passes through the external and the internal surfaces of the wall of the submerged part of the tuyere, Eq. (5.1), the densities of the flows on these surfaces q_m and q_w may differ significantly depending on the ratio between the areas of the surfaces F_1 and F_2, Fig. 5.5.

In the past, resulting from then existing technology of the tuyeres manufacturing, the design of the tuyeres for the open-hearth furnaces and the Berry-type tuyeres for the EAFs was such that the area of the surface F_1 considerably exceeded the area of the surface F_2. In these tuyeres, the concentration of the heat flows on the water-cooled surface occured, which resulted in the increase of the q_w in comparison with q_m by $F_1/F_2 = K$ times. The typical values of the coefficient of heat flow concentration K for both open-hearth furnaces and the Berry-type tuyeres were usually equal to 1.5–2.0, which led to the maximum values of q_w equal to about 3×10^3 W/m^2 in the slag and 10×10^3 W/m^2 in the metal.

In many instances, the heat flows of this high density could not be transferred to water by the regular convective heat transfer from the inner surface of the wall. At this surface, the local water boiling occurred in the boundary layer. The very mechanism of the heat transfer was altered, which deteriorated the durability of these tuyeres and resulted first in abandonment of the submerged blowing and then led to the wide-spread of the jet modules. Let us examine the conditions for the occurrence of local water boiling and the effect of this process on durability of the tuyeres.

5.3.2 Cooling of Tuyeres with Local Water Boiling

With increasing q_w, when α_w и t_w are constant, the temperature t_2 increases linearly and the lower the coefficient α_w, the quicker this increase is, formula (5.6), Fig. 5.7. However, such an increase is observed only to a defined limit. As soon as t_2 exceeds the water boiling temperature t_b at the given pressure, Table 5.1, by a few degrees, the further increase in t_2 slows down sharply. For example, when $\alpha = 15$ kW/(m^2 °C) and water pressure is $p_w = 8$ bar (abs), reduction of t_2 growth rate will take place at the temperature close to the boiling point $t_b = 170.5$ °C at this pressure (dotted broken line on the Fig. 5.7). This can be explained by occurrence and development of the local bubble boiling of water which significantly intensifies the heat transfer from the wall.

The mechanism of this process is as follows. When $t_2 > t_b$, despite the fact that water flow in general remains cold, the vapor bubbles are formed on the wall. When they detach from the wall, they pass through the boundary layer of water and condense when entering the cold core of the flow. These bubbles turbulize the laminar sub-layer of the boundary layer[2]; its thermal resistance to the heat flux

---

[2] See Chap. 3, Sect. 3.3.3.

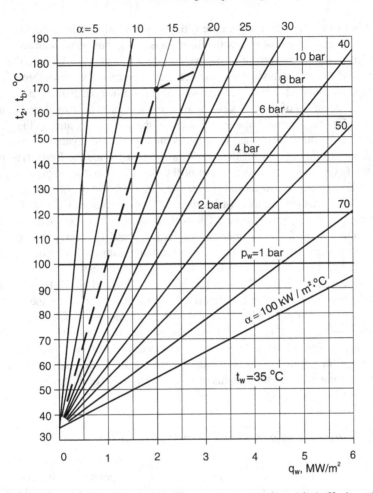

Fig. 5.7 Temperature of wall flown by water versus q_w, α and p_w (abs.). *Horizontal straight lines* are water boiling temperatures t_b at various values of p_w. The average water temperature is 35 °C

Table 5.1 Water boiling temperature t_b depending on pressure p, abs

p, bar	1	2	3	4	5	6	7	8	9	10
t_S (°C)	99.5	120.4	133.7	143.7	151.9	158.9	165.1	170.5	175.4	179.9

drops sharply, and the coefficient α_w increases respectively. As q_w grows further, the number of bubbles formed per unit time grows, and α_w continues to increase.

However, increase of α_w continues only to a certain heat flux level critical for the given conditions, $q_{w.cr}$. When the heat flux exceeds this level, the number of bubbles becomes so huge that they coalesce into a solid vapor film which has negligible thermal conductivity and completely insulates the wall from water. If the vapor film is stable, this leads to the abrupt growth of t_2 and t_1 to such high temperatures that heat transfer occurs through radiation only. In this case, the wall burns through immediately. Let us discuss the method of calculation of the wall temperature t_2 in the case of local boiling.

When the local water boiling is absent, a coefficient of convective heat transfer α_w in expression (5.6) depends mostly on the water velocity w. It can be calculated by the simplified formula (5.7) [2]:

$$\alpha_w = A \cdot \frac{w^{0.8}}{d_h^{0.2}}, \quad kW/\left(m^2\,{}^\circ C\right) \tag{5.7}$$

The coefficient A is determined by physical properties of water which depend on temperature. Its values are cited in Table 5.2.

d_h hydraulic diameter of a channel, m

For the round channels, $d_h = d$. For the channels of rectangle shape, $d_h = 4F/2(a + b)$, where F is cross-section area of the channel, a is the channel width, b is its height, and $2(a + b)$ is perimeter of the channel. For the relatively narrow ring and rectangular channels (slots) with the width of a, $d_h = 2a$.

When local boiling appears, convective heat transfer coefficient α_w increases to the value of α^*. The latter depends on two basic factors, namely the water flow velocity in the channel w and the intensity of turbulization of laminar sub-layer by vapor bubbles. The effects of both these factors are added together. The higher water velocity and the more intense boiling, the higher is α^*.

This mixed process of heat exchange is very complex. Various criterion formulas for determining α^* are known. These formulas summarize the results of vast experimental research. They are complex, contain the parameters which are hard to determine, and are not suitable for simplified calculations of the water-cooled tuyeres. For such calculations, it is recommended to determine the coefficient α^* depending on the ratio of the following two coefficients of heat transfer: coefficient α_b for boiling in large volume without forced water flow and purely convective coefficient α_w without local boiling which depends on flow velocity, formula (5.7). It is recommended to assume $\alpha^* = \alpha_w$ for $\alpha_b < 0.5\alpha_w$, $\alpha^* = \alpha_b$ for $\alpha_b > 2\alpha_w$, and it is recommended to use the interpolation dependence for the intermediate range $0.5\alpha_w < \alpha_b < 2\alpha_w$ [3]:

$$\alpha^* = \alpha \cdot \frac{4\alpha_w - \alpha_b}{5\alpha_w - \alpha_b} \tag{5.8}$$

Table 5.2 Dependence of the A-coefficient at formula (5.7) on water temperature

t (°C)	20	25	30	35	40	45	50	55	60
A	2.56	2.59	2.63	2.67	2.71	2.74	2.78	2.80	2.83

To determine α_b the well-known simplified formula for water boiling in a large volume can be used:

$$\alpha_b = 3.14 \cdot q^{0.7} \cdot p^{0.15}, W/(m^2 \,^\circ C) \qquad (5.8')$$

q heat flux density, W/m^2
p absolute water pressure, bar

In practice, the case of $\alpha_b < 2\alpha$ can be encountered only when water flow velocities w are very high. In most cases with local boiling, $\alpha^* = \alpha_b$ can be assumed. Since $\alpha^* > \alpha_w$, it can seem that the cooling of tuyeres in the local boiling mode can be effectively used for removal of large heat fluxes. However, in reality, this mode has significant drawbacks and cannot be recommended.

Firstly, when the local boiling occurs, the temperature t_2 cannot be below water boiling point t_b which exceeds 140–150 °C at the excessive water pressure in the element of 3–4 bar, Table 5.1. When the initial water temperature is about 25°C, the convective cooling mode allows assuring much lower temperatures t_2 and, respectively, improving the element durability, since the reduction of t_2 leads to the equal reduction of the external surface temperature t_1, formula (5.5).

Secondly, the local boiling mode is potentially dangerous due to the fact that under certain unfavourable conditions the bubble boiling in some local areas of the element can change to the film boiling which would immediately lead to burn-back. One should not count on the fact that when water velocity is quite high, about 3–5 m/s, the steady film boiling requires such high heat fluxes which are not achievable in the practical operation of EAFs. The research carried out on the transparent models of tuyeres with the aid of high-speed filming has shown that when configuration of the channels is complex, the high-frequency pulsations of water velocity occur in these channels. The water velocity can sometimes for the short periods of time drop to very low values in the poorly flown-around sections of the channels in turbulent zones behind the sharp corners, where the water flow detaches from the walls. In such zones the bubble boiling and the film boiling can continuously take turns. This does not lead to an immediate burn-back, but sharply increases the temperatures t_2 and t_1 and the wear rate. The specifics discussed above lead to the conclusion that in order to assure reliability and durability of the water-cooled tuyeres, the modes of operation with local boiling should be avoided. The cooling system should be designed in such a way that the temperature of the wall surface flown around by water t_2 is significantly lower that the boiling point t_b.

5.3.3 Jet Cooling

The density of the heat flow q_m from the melt to the submerged surface of the tuyere is distributed quite unevenly over this surface. A part of this surface where q_m reaches its maximum values is a critical zone of the tuyere. It is obvious that to avoid rapid wear and sharp shortening of the service life of the tuyere, its critical zone must be cooled most intensively. In the furnaces with flat bath, the tuyere cannot be submerged into the melt in the scrap charging zone. Therefore, the danger of the adverse effect of the oxygen jets reflected from the scrap pieces on the tuyere is eliminated. In this case, a part of the surface of the copper tip of the tuyere close to the oxygen nozzles is usually a critical zone.

To be efficient, the cooling system of the tuyere must be calculated so as to eliminate the possibility of quite undesirable local boiling of water in the boundary layer. For this reason, the cooling mode of the critical zone in case of local boiling is not further examined. In the absence of local boiling, the cooling intensity is defined by the coefficient of convective heat transfer α_w. In order to increase this coefficient, it is necessary to increase the speed of water w, Eq. (5.7). With that, the problem of the required water pressure appears, since, with increase of w, both the hydraulic resistance and corresponding required water pressure increase directly proportional to w^2.

The hydraulic resistance is equal to a difference in the water pressure at the entrance and at the exit of the tuyere: $\Delta P_w = P_1 - P_2$. In furnace operation, this pressure difference is always invariably limited by the water supply system of a given shop. In order to supply water to mobile tuyeres, the supply and drain collectors were used in most cases. These collectors are present in every furnace and supply water to the wall and roof panels as well as to the other water-cooled elements. However, the pressure difference between these collectors usually does not exceed 2.0–2.5 bars, which is insufficient for the effective cooling of the mobile tuyeres, especially of the open-hearth-type and the Berry-type tuyeres in case of the submerged blowing. This was one of the causes of abandoning the usage of these tuyeres due to their low durability.

So far, installing the special pumps for supplying high pressure water to mobile tuyeres was avoided, since it involves additional costs growing rapidly with an increase of pressure. Hence, the designers of the mobile tuyeres for submerged blowing always encounter the problem of ensuring reliable cooling of the critical zone with the minimal hydraulic resistance of the entire tuyere. In this problem, the minimum required water flow rate per tuyere V_w is a preset initial value which can be calculated using the formula (5.9) with the error not exceeding 1 % [2]:

$$V_w = Q/1.15 \cdot \Delta t_w, m^3/h \qquad (5.9)$$

Q   maximum heat flow absorbed by water, kW
Δt_w allowable increase of water temperature in the tuyere, °C

When chemically unprocessed water is used, the value Δt_w should be selected so that the final water temperature at the tuyere exit does not exceed 45 °C to avoid salts precipitation. In any case, increasing Δt_w raises the temperatures t_1 and t_2 and increases probability of local boiling occurrence. That is why it is usually assumed in the calculations that the difference Δt_w should not exceed 30 °C.

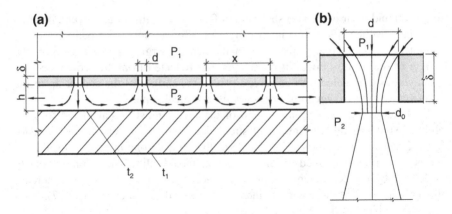

Fig. 5.8 Schematic diagram of jet cooling (designations are given in the text)

The formula (5.7) relates to the flow of water along the cooled walls, for example, to the flows in the pipes. In this case, α_w increases mainly due to the decrease of the thickness of the laminar sublayer which is caused by the increasing in value transverse turbulent fluctuations of the velocity of the flow. Another more effective method of increasing α is known, when high-speed jets of the cooling water or gas are directed onto the wall at a right angle. This method is widely used in practice, for instance, for cooling the steel sheets after rolling. The schematics diagram of such jet cooling of the water-cooled tuyeres and other furnace elements are shown in the Fig. 5.8.

The jets leaving the openings at high speed hit the cooled surface and spread over it in all directions. In the impact zone, the boundary layer and its laminar sub-layer are being completely destroyed, which results in very high values of α. In the flat jets spreading over the surface, the coefficient α drops rapidly with distance from the impact zone. That is why the average values of α W/(m² °C) strongly depend on the pitch between the jets in both the transversal and axial directions. These average values of α are determined by the formula (5.10):

$$\alpha = 0.01 \cdot \frac{2\lambda}{x} \cdot Re^{0.75} \cdot \frac{(9-0.45x)/d}{(h/d)^{0.2}} \cdot \frac{Pr^{0.33}}{0.88} \qquad (5.10)$$

x	pitch between the jets, m
λ	thermal conductivity of water, W/(m °C)
$Re = w_0 \cdot d/v$	the Reynolds number, where w_0 is initial velocity of jets, m/s, d is diameter of the openings, m, and v is kinematic viscosity of water, m²/s
h	length of the jets from exit of the opening to the cooled surface, m
Pr	the Prandtl number

The dependence between α and x calculated by the Eq. (5.10) is shown on the Fig. 5.9. According to this equation, coefficient α also strongly depends on the Reynolds number, and consequently, on the initial velocity of jets w_0 and the

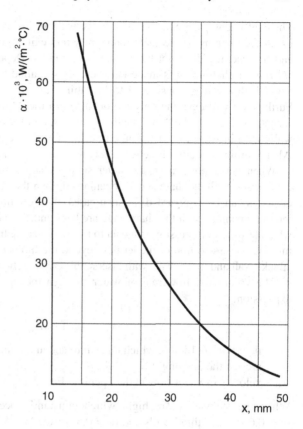

Fig. 5.9 Heat transfer coefficient α versus pitch between jets, x

diameter of the openings d. The relative length of the jets h/d has a very weak effect, since its degree in the formula (5.10) is 0.2.

The formula (5.10) has been obtained in the experiments with free jets, i.e., with the jets which spread after impact in free space. In our case, the jets are not free, since they spread in the cooling chamber limited by the wall with openings and the cooled wall. However, the formula (5.10) can be used for this case as well under condition that the average velocity of the flow moving along the cooled surface towards the exit from the chamber is quite low in comparison with the initial velocity of the jets w_0. This velocity is determined by the expression:

$$w_0 = \varphi(2 \cdot \Delta P / \rho_w)^{0.5} \qquad (5.11)$$

φ velocity coefficient which takes into account the hydraulic resistance of the opening

$\Delta P = P_1 - P_2$ pressure difference between the water supply chamber and the jet cooling chamber, Pa

ρ water density, kg/m^3

The coefficient φ depends on the relative wall thickness δ/d and on the surface finish of the opening. It means that the entry and exit edges of the opening are sharp and the opening does not have machining marks. Depending on the value of the δ/d ratio, a distinction is made between the so-called "thin" and "thick" walls. If $\delta < 2d$, the wall is considered to be "thin", and if $\delta > 4d$, the wall is "thick". Further, we will examine only the openings in the "thin" walls which are optimal for jet cooling. For the "thin" walls, $\varphi = 0.98$, and for the "thick" walls, or for the walls with inserts, $\varphi = 0.8$. Thus, in case of $\delta < 2d$ and the preassigned value of ΔP, the initial velocity of the water jets w_0 is maximum possible.[3]

When water flows from the water supply chamber of sufficiently large size in comparison with the diameters of openings d, then the velocities of water before the openings can be disregarded. In such cases, the flow lines of water at the entrance into the openings with the sharp edge are bent under the action of the inertial forces, which leads to compression of the jet to the diameter d_0. In case of the "thin" wall, the minimum cross-section of the jet lies beyond the limits of the wall, Fig. 5.8b. In the "thick" wall and in the walls with inserts, this cross-section lies inside the opening.

The volumetric flow rate of water through the opening is determined by the expression:

$$v_w = \mu \cdot f \cdot w_0 \tag{5.12}$$

μ flow rate coefficient which takes into account compression of the jet
f area of the opening, $f = \pi d^2/4$
w_0 velocity of water flow at the opening, Eq. (5.11)

When the Re values are high, which constantly occurs in real practice, then $\mu = 0.6$ for the "thin" walls and $\mu = 0.8$ for the "thick" walls. Thus, the flow rate coefficient μ has minimum value in case of the "thin" walls.

The possibility to alter the intensity of cooling of different sections of the water-cooled element by varying the pitch between the jets x is the fundamental advantage of jet cooling. This allows to concentrate in the critical zone of the tuyere the largest number of the jets per unit area of the cooled surface and to gradually reduce the intensity of cooling, increasing pitch x in proportion to distance from the critical zone. None of the other cooling methods has these options. However, to realize this advantage, there must be sufficiently large number of the jets or, what is the same, sufficiently large number of openings in the wall separating the water supply chamber and the cooling chamber.

The possible number of the openings n with diameter d correlates unambiguously to the preassigned values of the flow rate of water per tuyere V_w and initial velocity of jets w_0, Eq. (5.12).

$$n = V_w/\mu \cdot f \cdot w_0 \tag{5.13}$$

$f = \pi d^2/4$ area of one opening, m^2

---

[3] The same can be said about the other incompressible liquids and gases.

As it was already noted, in order to achieve maximum values of α, it is necessary to use the maximum possible share of the available water pressure difference for attaining the maximum values of w_0 in the critical zone, Eq. (5.11). For this purpose, the hydraulic resistance of all other sections of the water circuit of the tuyere must be reduced to a possible minimum. In case of preassigned values of V_w and w_0, the number of the jets n increases sharply with the decrease in the diameter of openings d. The pitch between the jets decreases directly proportional to the increase in n per unit area of the cooled surface. The analysis of the Eq. (5.10) and Fig. 5.9 shows, that this is accompanied by an increase in α to the values which are practically unattainable in case of the longitudinal flow of water along the cooled surfaces, because required values of velocity and hydraulic resistance are too high. This allows ensure the high durability of the tuyeres with the jet cooling even under the severe conditions of the submerged blowing in case of the extreme values of the heat flows.

5.4 Roof Oxygen Tuyeres Increasing the Rate of Scrap Melting

5.4.1 Layout of the Tuyeres in the Furnace, Their Design and Basic Parameters

In order to considerably accelerate scrap melting with the help of the tuyeres, they must be submerged into bath close to the charging zone, in much the same manner as shown in Fig. 5.10. The oxygen jets of the tuyeres (1) directed towards the zone create the co-current flows of the melt, which increase the intensity of mixing in the charging zone as well as the velocities of the liquid metal flowing around pieces of scrap. In the same figure a jet module (2) is shown as well. Obviously that the effect of an oxygen jet of the module on metal stirring in the charging zone is very weak and does not compare with that of jets of the submerged tuyere (1) because of a large distance between the module and the bath surface.

Different mechanisms can be used for lowering and lifting the tuyeres (1), including mechanisms similar to those used for moving electrodes. These mechanisms as well as the arms of tuyeres must have a sufficient mechanical rigidity ensuring that the tuyeres move along the axis of the openings in the roof without contacting the walls when the furnace is tilted. This is essential for successful operating of the system controlling the position of tuyeres relative to slag-metal interface, Sect. 5.4.2.

A drawback of the roof tuyeres is their large length. It increases the required flow rate of cooling water and makes the design of the mechanisms more complex due to large stroke of the tuyeres which must be completely brought out of the furnace freeboard when the roof is lifted and swung away. In the shaft furnaces of the COSS-type, charging a part of the scrap from the top, which requires roof

Fig. 5.10 Positioning roof
tuyere (1) on a furnace in
comparison with jet module
(2)

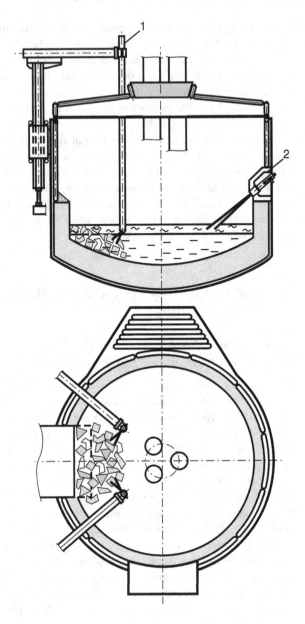

swinging, is carried out during the first heat of each of the long series of heats.
However, it should be taken into consideration, that immediate roof swinging
might become necessary at any moment of the process due to occurrence of an
unpredictable emergency situation.

Both the length and stroke of the tuyeres decrease with the decrease of the
height of the roof above the bath level. This decrease is reasonable, since it

reduces the heat losses by the furnace freeboard. According to the design considerations, the height of the roof can be significantly reduced in the furnaces with flat bath, Chap. 1, Sect. 1.3.7. Bath splashing during oxygen blowing is one of the basic factors limiting this possibility. The lower the roof, the more metal and slag splashes hit the central ceramic part of the roof. They saturate ceramics with iron oxides, which sharply decreases its refractory properties, accelerates wear, and shortens the service life of the roof. Therefore, in order to significantly decrease the height of the freeboard, it is necessary to sharply reduce the intensity of splashing. Submerged blowing of the bath ensures such reduction, Sect. 5.2.3.

Figure 5.11 shows schematic diagram of a design of a tip of three-nozzle oxygen tuyere which uses advantages of jet cooling. The tuyere is designed for submerging into the bath to the slag-metal interface near the scrap charging zone. The tip has a spherical shape which is characterized by the minimum surface area. This is very important for the decrease in the required number of the cooling jets. The tip has water supply chamber (1) and the jet cooling chamber (2). These chambers are separated by the wall (3) with openings (4) forming the water jets which are directed towards the spherical wall of the tuyere at the 90° angle (5). The closer to the critical zone, i.e. the section of the surface with the nozzles, the smaller the pitch between the openings.

At the exit from the jet cooling chamber, the partition wall (3) narrows down the external ring-shaped channel (6) through which water is removed from the tuyere. This is necessary for increasing the velocity of water and the intensity of cooling of the part of the external pipe which can be submerged in the foamed slag. Above, outside the slag zone, where the density of the heat flows affecting the tuyere sharply decreases, the ring-shaped channel (6) is wider, which reduces the velocity of water through the entire remaining length of the tuyere as well as its general hydraulic resistance.

The oxygen is delivered to the tuyeres through the central pipe (7). Three lateral supersonic oxygen nozzles are tilted at different angles with regards to the vertical. The tuyere is positioned in the holder arm in such a way that all the nozzles are directed toward the scrap charging zone. Thus, the device shown in the Fig. 5.11 satisfies all the stated above basic principles of designing of the highly durable oxygen tuyeres ensuring an increase in scrap melting rate.

The external spherical wall of the tip and the part of pipe (7) which can be submerged in the slag are made of high-quality copper; the partition wall with openings (4) which require fine surface finish must be made of stainless steel. Other pipes are made of steel. For compensating the thermal expansion of the external pipe on the upper end of tuyere, the expansion joints are installed. The key parameters of the tuyere for the shaft furnace with capacity 170 t and tapping weight of 125 t can be determined by calculation.

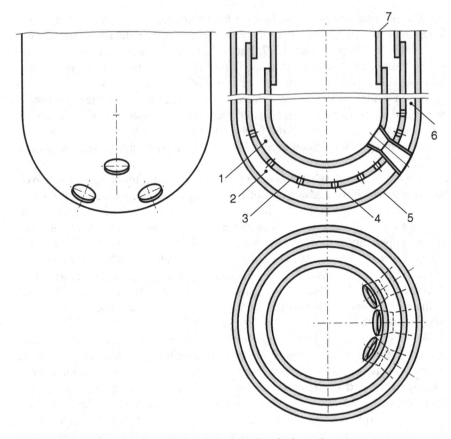

Fig. 5.11 Tip of oxygen roof tuyere (designation are given in the text)

5.4.1.1 Calculation

Oxygen Flow Rate V_{O_2}

Let us assume that the specific intensity of oxygen blowing J is close to its maximal values used in practice, i.e., J = 1.0 m^3 (s.t.p.)/t min; hot heel weight is 40 t; average weight of liquid metal in the furnace is (40 + 60) = 100 t; quantity of tuyeres—2; oxygen flow rate per one tuyere: $V_{O_2} = 1.0 \cdot \frac{100}{2} = 50$ m^3/min (3000 m^3/h); quantity of nozzles—3. At oxygen pressure in front of the nozzles equal to 12 bar, the critical cross-section diameter of a nozzle d_{cr} is equal to 15 mm.

Water Flow Rate Per One Tuyere V_w

The length of the tuyere in the furnace, considering a lowered height of the roof, amounts to 2.85 m including 0.35 m being submerged into the foamed slag. External diameter of the tuyere is 0.127 m. Densities of heat fluxes to the tuyere are 0.8 MW/m^2 in the freeboard and 1.4 MW/m^2 in the slag. The total heat flux to the tuyere is: $Q = 1.4 (\pi \cdot 0.127 \cdot 0.35) + 0.8 (\pi \cdot 0.127 \cdot 2.50) = 0.992$ MW or 992 kW.

The minimum required water flow rate is calculated by formula (5.9): $V_w = Q/1.15 \cdot \Delta t_w$. Assuming that Δt_w is equal to 30 °C we find: $V_w = 992/1.15 \cdot 30 = 28.7$ m^3/h or 0.0080 m^3/s.

Initial Velocity of Water Jets w_o

Let us assume that pressure difference between a chamber of water supply (1) and jet cooling chamber (2), Fig. 5.11, amounts to 3.5 bar (350,000 Pa). w_o is calculated by formula (5.11): $w_o = \varphi (2 \cdot \Delta P/\rho_w)^{0.5}$. At 20 °C, $\rho_w = 998$ kg/m^3, we find $w_o = 0.98 (2 \times 350,000/998)^{0.5} = 25.9$ m/s.

Quantity of Jets n and Pitch of Them x

By the clogging conditions a minimal diameter of openings in the wall (3) d is assumed to be equal to 3.5 mm. The area of each opening $f = 0.0000159$ m^2. The number of openings $n = V_w/\mu \cdot f \cdot w_o$. For openings in the thin wall a discharge coefficient μ is equal to 0.6. Using all known data we can determine $n = 0.008/0.6 \times 0.0000159 \times 25.9 = 32$.

The area of the semi-sphere surface cooled by the jets is the F. At the thickness of external wall of the tuyere equal to 8.5 mm radius of this surface R amount to 55 mm, and then $F = 2\pi \times 55^2 = 19,000$ mm^2. The area cooled by one jet is $19,000/32 = 594$ mm^2; and the pitch of openings is $x = 594^{0.5} = 24.4$ mm.

Heat Transfer Coefficient α_w

Dependence of the coefficient α_w on the jets pitch x calculated according to formula (5.10) is shown in Fig. 5.9. At the average value of $x = 24.4$ mm the heat transfer coefficient α_w is equal to 35.1 kW (m^2 °C). The optimal distribution of jets over the cooled surface allows decreasing the pitch x in a danger zone close to oxygen nozzles to 15 mm which increases the α_w up to 65.2 $kW/(m^2$ °C).

Temperatures of The Tuyere Tip Wall in Danger Zone

The external surface of this wall has a temperature of t_1, and a temperature of internal surface flown by water is t_2. Since the thickness of the wall is small as compared with its radius densities of the heat flux on both surfaces can assume to be identical and equal to 1400 kW/m^2 when the tuyeres are submerged into the slag. The temperature t_2 is determined by the known formula:

$$t_2 = t_w + q_{sl}/\alpha_w \qquad (5.14)$$

Let us assume that a water temperature at the jet cooling area t_w is equal to 20 °C because it does not practically differ from a temperature at the tuyere inlet. At the $t_w = 20$ °C, $q_{sl} = 1400$ кВт/м^2, and $\alpha_w = 65.2$ kW/(m^2 °C), the temperature t_2 will be equal to: $20 + 1400/65.2 = 41$ °C.

When contacting the tuyere tip with liquid metal the q_m will be equal to: 4500 kWt/m^2 and, consequently, $t_2 = 20 + 4500/65.2 = 89$ °C.

Thus, even in the case of contact with the liquid metal an occurrence of local boiling in the danger zone of the tip is eliminated.

Temperature of the surface of the tuyere tip is determined by the known formula:

$$t_1 = t_2 + q_m \cdot \delta / \lambda \qquad (5.15)$$

δ 0.0085 is a thickness of the copper wall, m
λ the coefficient of thermal conductivity of copper depending on an average temperature of the wall, Fig. 5.6

The value of the λ is resulted from the preliminary calculations of the t_1. The temperature $t_1 = 41 + 1400 \times 0.0085/0.39 = 72$ °C when the tip is in the slag and $t_1 = 41 + 4500 \times 0.0085/0.38 = 142$ °C in the liquid metal, respectively.

Such low temperatures of the surface of the most heat-stressed section of the tip not only in case of submerging in slag, but also in case of its contact with metal ensure quite prolonged service life of the tuyere, which allows extensive use of the oxygen submerged blowing method as most effective not only in the furnaces with the flat bath, but in other EAFs as well.

The total calculated hydraulic resistance of the tuyere is approximately 4 bar, and losses of pressure in the zone of jet cooling comprise 87 % of this value. In order to exclude the possibility of blocking the openings with a diameter of 4.5 mm, the preferable water supply system for the tips of tuyeres is a system with filters and pumps designed for the pressure of approximately 6 bar.

5.4.2 Controlling Position of Tuyeres at the Slag-Metal Interface

The slag-metal interface level is rising constantly in the course of scrap melting. In the 125 ton furnace this rise is about 700 mm over the entire scrap melting time.

To maintain position of the tip near the slag-metal interface, the tuyere must be constantly raised. This requires the use of an automated control system receiving data on the position of the tip of tuyere relative to the metal surface.

It is necessary to consider that submerging of tuyere in metal even as much as by 50–80 mm is undesirable. The experiments carried out on the open-hearth furnaces demonstrated that such submerging does not increase the effectiveness of the blowing, but expose the tuyere to the extreme thermal loads, which adversely affects its durability. The level of the metal can differ significantly from heat to heat depending on the mass of hot heel and the wear of the furnace bottom. Therefore, programmed control is hardly suitable in this case. The sensor determining position of the tuyere tip relatively to the metal surface with sufficiently high accuracy is required.

As a result of the persistent attempts to solve this problem for the open-hearth furnaces, the sensors using different principles of operation have been tested [1]. They can be divided into two groups. The first group are the sensors built into tuyere. They include pneumatic, radiation, temperature, vibration, and other sensors. They all proved to be insufficiently reliable. Furthermore, they all make the tuyere design and its manufacturing technology more complicated. For these reasons the built-in sensors did not find a practical use.

The sensors of the second group are installed outside of the tuyere. They include the sensors which detect the tuyere weight loss, change in the cooling water temperature, change in the electromotive force generated in the tuyere or electrical resistance between the electrically isolated tuyere and the furnace shell, and others. Out of all of these, only resistance sensors have accuracy and reliability required for the use in the electric arc furnaces.

A typical change in the electrical resistance of tuyere in relation to its position in the furnace is shown in Fig. 5.12 [1]. When the tip of tuyere is positioned near the furnace roof, the resistance of the tuyere is 25–30 kΩ. As the tuyere is lowered down, the resistance drops sharply reaching the value of several ohms at the moment when it contacts the slag. In case of further submerging into the slag, the resistance decreases considerably slower, and remains practically constant after the tuyere reaches the metal. Inflection points a and b on the graph shown in Fig. 5.12 correspond to the position of the tuyere on the phase interfaces: gas–slag and slag–metal.

It should be noted, that the absolute value of resistance does not allow determining the position of the tuyere in the phases of the freeboard. The conductivity of the gas phase is determined by its ionization which depends on the temperature and a number of other factors. The same relates to the slag layer located between the tip of the tuyere and the metal. Its resistance depends not only on the thickness of this layer, but also on the temperature, composition and consistency of slag which are changing in the course of the heat.

After the tip gets in contact with the metal, the current passes through the circuit consisting of the metal and inner layer of lining. Because of the large cross section and low specific resistance of these conductors, the resistance of the tuyere maintains approximately constant small value regardless of its position in the metal. When the tip of the tuyere is positioned in the atmosphere of the furnace and in the layer of slag, the resistance of tuyere fluctuates; in the metal the fluctuations stop. Such changes in the tuyere resistance allow determining the moment

Fig. 5.12 Electrical tuyere resistance R versus tuyere position. Range of pulsations is shown by *dashed lines*. Points "a" and "b" are interfaces of phases

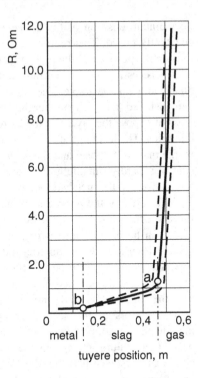

when the tip gets in contact with the metal surface. When lowering down of the tuyere the minimum absolute value of resistance and stoppage of the fluctuations correspond just to the moment of its contacting the metal.

Automated control systems with the electrical resistance sensors were successfully used for controlling the position of tuyeres relative to slag-metal interface in many open-hearth furnaces. By applying these sensors it was possible to keep the tips of the tuyeres near the metal surface with the 20 mm accuracy. This high accuracy is achieved due to the fact that during the blowing in the lower position of the tuyere its tip gets into contact with the metal [1]. Similar systems should also be used in the shaft furnaces with the flat bath.

References

1. Markov BL (1975) Methods of open-hearth bath blowing. Moscow, Metallurgy
2. Toulouevski Y, Zinurov I Innovation in electric arc furnaces. scientific basis for selection. 2nd edn (revised and supplemented). Springer
3. Malek AO (2004) DC EAF with DRI feeding rates through multipoint injection. MPT Int 2:58–67
4. Rondi M, Bosi P, Memoli F (2002) New electrical and chemical technologies implemented in the dalmine steel plant. MPT Int 5:44–51

5. Mukhopaahyay A, Coughlan R, Ometto M et al (2012) An advanced EAF optimization suite for 420-t jumbo DC furnace at Tokyo sreel using danieli technology. In: Proceedings of 10th European electric steelmaking conference, Graz, Austria
6. Pujadas A, McCauly J, Tada Y et al (2002) Electric arc furnace energy optimization at nucor yamato steel, In: 7th European electric steelmaking conference, Venice, May, 2002

Chapter 6
Fuel Arc Furnace FAF with Flat Bath: Steel Melting Aggregate of the Future

Abstract The systems of scrap preheating and submerged oxygen blowing in combination with a device increasing volume of the scrap charging zone allow offering the steel melting aggregate of new type, i.e., fuel arc furnace FAF. A base of this aggregate is a shaft furnace with scrap pushers. The predominance of fuel energy in the heat balance and ultra-high power of burner devices exceeding transformer power are fundamental features of the aggregate. The given designed analysis of basic performances of the new aggregate carried out with use of the reliable experimental data on melting scrap samples in liquid metal shows that the aggregate FAF of 185 t capacity with tapping weight of 130 t, equipped with a 120 MBA transformer and 94 MW burner devices power, at 100 % scrap content in a charge preheated up to an average mass temperature of 800 °C, will have productivity of 312 t/h at electrical energy consumption of 214 kWh/t and natural gas flow rate of 17.5 m^3/t. Therewith all the advantages of furnace operation with flat bath are kept; high economic efficiency and complete satisfying with environment regularities are ensured. Such performances of the FAF exceed considerably those of the best modern electric arc furnaces of similar capacity. Replacement of the EAF by FAF will contribute to an increase of the share of electric steel in total production. This will improve effectiveness of the production especially its ecological characteristics.

Keywords Fuel arc furnace FAF · Fundamental features · Intensive replacement of electrical energy with gas fuel · Ultra-high heat power of burner devices · Design furnace operation performances · Advantages · Prospects · Replacement of EAF by FAF

6.1 Design and Operation

The content of the previous chapters makes it possible to describe in sufficient detail the steel melting aggregate capable of replacing modern EAFs processing scrap. Basic design of this aggregate is a shaft furnace with continuous charging of

© The Author(s) 2015

Y.N. Toulouevski and I.Y. Zinurov, *Electric Arc Furnace with Flat Bath*, SpringerBriefs in Applied Sciences and Technology, DOI 10.1007/978-3-319-15886-0_6 preheated scrap into liquid bath. To successfully compete with the modern EAFs of the same capacity, the new aggregate must have considerably higher productivity and considerably lower electric energy consumption.

Increasing productivity requires a sharp reduction of duration of scrap melting in liquid metal which until now has been a bottleneck of the process on the furnaces with flat bath. To considerably reduce electric energy consumption extensive substitution of electric energy with energy of fuel is required. In the new aggregate, these objectives are achieved due to two basic innovations, i.e. preheating of scrap in the shaft to 800 °C by high-power gas-oxygen burner devices as well as directed toward the scrap charging zone oxygen blowing using the roof tuyeres submerged into the melt to the level of the slag-metal interface. The energy of the fuel is dominant in the heat balance of this aggregate. This justifies its name—fuel arc furnace (FAF).

Let us review basic design and technological features of the FAF. The freeboard and the bath of the furnace are asymmetric. They are similar to ellipse-shaped from the side of the scrap charging window. This allows to increase the width of the window and the volume of scrap charging zone. Increasing the depth of the bath due to increase of the hot heel mass serves the same purpose.

The operational experience of the Consteel furnaces demonstrated that on the furnaces with flat bath, increasing hot heel mass to 40–50 % of the mass at the tapping considerably reduces the duration of scrap melting in liquid metal [1]. This is explained by an increase in depth and volume of scrap charging zone as well as by the fact that that scrap melting begins at a higher temperature of the metal which, due to the short power-off time of the heat, does not have time to cool down after tapping.

A special charging device is used to increase the volume of charging zone and to distribute scrap evenly. Its key part is a water-cooled plate (1) which can be moved quite far into the freeboard of the furnace using a hydraulic cylinder (2), Fig. 6.1. The width of the plate corresponds to the width of the window for scrap charging into the bath.

The movements of the plate (1) and the scrap pusher (3) are coordinated with each other. The plate starts to move into the freeboard at the same time as the pusher starts to move discharging scrap from the shaft. The stroke of the plate is considerably shorter than that of the pusher. Therefore, the plate moves slower so that both the plate and the pusher reach their end positions and stop at the same moment of time. In the process of discharging from the shaft, the scrap pieces fall onto the moving plate and then falls off its front and side edges into the bath. Thus, the uniform distribution of the scrap throughout the entire area of the charging zone is achieved. Then the plate and the pusher return to the initial positions. As it occurs, the scrap remaining on the plate is uniformly pushed into the bath using the stopper (4). All following portions of scrap are discharged from the shaft in the same manner.

Since returning back to the initial positions takes very short time, the portions of scrap following each other are charged into the bath practically continuously. As a matter of fact, such charging does not differ from the conveyor charging on

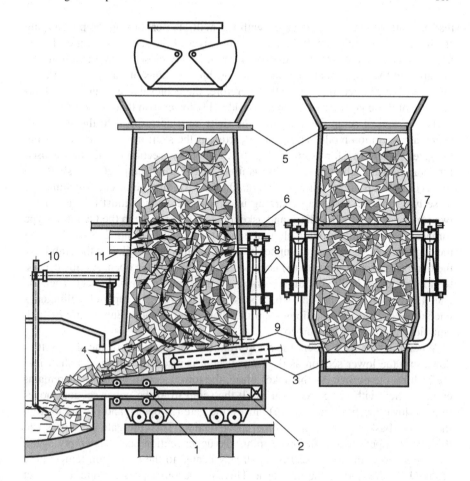

Fig. 6.1 Fuel arc furnace FAF (designations are given in the text)

the Consteel furnaces. One of the major advantages of such charging is that it allows operating the burners in the constant power regime. This stationary regime is the most efficient since it makes it possible to considerably simplify the design of the devices as well as increases the reliability of their operation and simplifies aggregate operation control.

With each portion, the pusher discharges from the shaft the lowest layers of scrap heated to the highest temperature. If this temperature is consistent with the maximum permissible temperature, then, in case of a constant power of the burner devices providing the heating, the risk of scrap overheating, its excessive oxidation, and yield reduction is eliminated. As soon as each portion of scrap reaches the maximum permissible temperature, it is discharged from the shaft immediately. Such stationary operating mode of the shaft scrap heater when the scrap is charged into the furnace bath at a constant, maximum permissible temperature and the burner devices operate at a constant maximum power is optimal. Let us note that the bath of the FAF is charged with the multiple portions of scrap of a quite small mass. Due to the small thickness of the layer of the scrap forming each portion, the average-mass temperature of a portion is very close to the maximum permissible. In the finger shaft furnaces, two or three portions of many times greater thickness are charged into the bath. Therefore, the achievable average-mass temperature of these portions of scrap is considerably lower than that of the FAF.

The system of scrap charging into the shaft must correspond to the stationary aggregate operation mode. This system partitions the shaft into two sections using the gates (5) and (6), for example, Fig. 6.1.[1] The lower section of the shaft is used for scrap heating. The upper section is used for charging scrap into the shaft and for feeding scrap into the lower section. On the FAF, in case of equal volumes of these two sections, the scrap charging into the upper section must be carried out two times per heat by the charging basket or, as it takes place in the Quantum-type furnaces, by the chute moving along the inclined ramp [3], Chap. 2, Sect. 2.3.1. At the beginning of each heat, the upper empty section is being filled with scrap from the first basket when the gate (6) is closed and the gate (5) is open. As this occurs, the lower section is filled with the scrap heated during the previous heat. Then the gate (5) closes, and the gate (6) opens. Such charging completely eliminates uncontrolled dust-gas emissions from the shaft into the shop. Entrapping of such emissions requires significant costs.

When the gate (6) is open, the scrap from the upper section of the shaft settles down into the lower section as the heated scrap is discharged from the shaft into the bath by the pusher. As this takes place, the lower section of the shaft remains entirely filled with scrap. As soon as all the scrap from the upper section moves into the lower section, the gate (6) closes, the gate (5) opens, and the scrap from the second basket is charged into the upper section. Then the gate (5) closes and the gate (6) opens again. Before tapping, the upper section becomes empty again, and the scrap in the lower section is already heated to the maximum temperature required for the start of the next heat. This also becomes possible due to the fact that scrap heating can still carry on for the short period of time after the pusher operation stops. Constant presence of the hot scrap in the lower section ensures the reliable ignition of combustible mixture and its complete combustion in the layer of scrap.

Since the upper pipes (7) of the burner devices (8) are located near the gate (6) and the lower pipes (9) are close to the pusher (3), Fig. 6.1, the combustion products pass through the scrap layer of a constant thickness. This ensures the required reduction of the temperature of the combustion products at the entrance into the burner devices as well as the steadiness of this temperature throughout the entire period of scrap heating. The latter is also one of the necessary conditions of the stationary operating mode of the REF-system.

The important feature of the shaft heater of the scrap of the fuel-arc furnace is its aerodynamics which is significantly different from aerodynamics of scrap

---

[1] It is also possible to use for this purpose the EPC charging system, Chap. 2, paragraph 2.3.2 [2].

heating by the off-gases. Due to the very high power of the burner devices as well as to recirculation of the combustion products, the amount of gases passing through the scrap layer per unit time in the FAF exceeds the flow of the off-gases by more than twice. In addition, the velocity of the gases passing through the layer of scrap, the uniformity of their distribution over the shaft cross-section, and the intensity of the heat transfer processes are much higher than those of the shaft furnaces which use the off-gases for scrap heating. All this contributes to reaching higher average-mass temperature of the portions of scrap being charged.

The above-noted advantages are ensured due to the following aerodynamic key features of the shaft of the FAF. The flow of the gases in the upper part of the lower section of the shaft is split, Fig. 6.1. Its smaller part equal to an amount of combustion products formed during the combustion of the gas fuel in oxygen is removed from the shaft through the gas duct (11) and is directed for purification. The larger part of the flow is drawn into the burner devices by the oxygen injectors through the pipes (7) and then blown back into the shaft through the pipes (9). This part of the total flow of the gases circulates continually through the closed loop, i.e. the burner devices—the layer of scrap in the lower section of the shaft—the burner devices. The amount of circulating gases is defined by the power of oxygen injectors which depends on the heat power of the burner devices.

The hydraulic regime of the shaft scrap heater of the fuel arc furnace is of great significance for its effective operation. The intensity of suction of the gases through the gas duct (11) must be maintained at the level such that, in case of the positive pressure of the gases under the closed gate (5), the average pressure at the scrap discharging window is close to the pressure at the adjacent window in the side wall of the furnace. This hydraulic operating mode of the shaft heater is also optimal for the entire aggregate including the furnace, since it enables maintaining exchange of the gases between the shaft and the furnace freeboard at the minimum level.

Sealing the gap between the windows is of paramount importance, since infiltration of the significant volumes of air both into the furnace freeboard and into the shaft decreases the effectiveness of the aggregate operation. The air curtains and other technical means can ensure the sharp reduction in air infiltration.

The average pressure at the level of the side wall window of the furnace depends on the pressure under the roof. Therefore, to maintain the optimum hydraulic operating mode of both the furnace itself and the shaft scrap heater, the pressure under the roof must be maintained at the strictly defined level using automated control. The sensors needed for this have been developed and are successfully used in the systems of automated pressure control in the EAFs. One of the most reliable sensors is the sensor supplied by the Preisman Technology Inc., Canada [4].

Another feature of the fuel arc furnace FAF is the reduced height of the freeboard. This allows to decrease the length of the roof tuyeres, the water flow rates, and the heat loss with water cooling side-wall panels as well as the oxygen tuyeres (10).

Basic performances of the FAF operation were determined using computational methods for the 185-t furnace with the tapping weight of 130 ton.

6.2 Design Furnace Operation Performances

6.2.1 Hourly Productivity

As it has been already emphasized, the productivity of operating furnaces with the flat bath is mainly restricted by the insufficient rate of melting a scrap within a limited volume of the liquid metal. It is this fact which is a bottleneck of the entire process. In order to define a potential level of FAF's productivity it is necessary to estimate maximal scrap melting rates which can be achieved in practice due to high-temperature scrap preheating and improving means of oxygen bath blowing as well. For such estimation the well-known experimental data obtained by the method of an immersion of cylindrical samples in the crucible of induction furnace containing liquid metal were used, Chap. 3 [5]. When immersing of samples the furnace was switched off and so the samples were being melted under conditions of free (natural) convection without forced metal stirring. The data of the work [5] were used at all the steps of calculations. These data have allowed determining a design hourly productivity of the FAF with accuracy sufficient for practical conclusions.

The calculations given below have required substituting a real scrap with its equivalent model. Let us notice, that there is no calculation of scrap melting rates (which is of interest to practice) which can be done without such a substitution. The scrap comprised of steel cylinders d in diameter and L in length was taken as the equivalent model. This model is most convenient for comparison of the results of calculations with the experimental data obtained by melting of the cylindrical samples.

With regard to the process of heat transfer from liquid metal to scrap, the ratio of the total surface area of scrap pieces F to their overall mass M is the key characteristic of scrap. Under otherwise equal conditions, the higher the ratio F/M, m^2/t, the shorter the melting time. Therefore, it can be assumed that the equality of the ratios F/M is a necessary condition of similarity of the real scrap to the equivalent scrap comprised of identical cylindrical pieces.

The calculations have shown that, despite a considerable spread of data on granulometric composition of scrap used in varied plants, the F/M equal to approximately 20 m^2/t can be assumed as the typical average value for scrap with volume density of 0.7 t/m^3. Cylindrical pieces of dimensions $d = 0.025$ m and $L = 0.4$ m correspond to this ratio. The actual scrap usually contains a lot of similar pieces as well. The volume of such a piece amounts to 0.196×10^{-3} m^3, the mass is 1.55 kg, and the surface area (without considering fronts surface area) is 0.0314 m^2. The ratio F/M $\cong$ 20 m^2/t is right both in an individual piece of scrap $(31.4/1.55 = 20.2 \ m^2/t)$ and in all n pieces of scrap located in a charging zone of a volume V_z.

An expression for the volume of all scrap pieces in the zone V_{st} is derived from the formula (3.16) determining the porosity of the zone P, see Chap. 3, Sect. 3.4.3. So, $V_{st} = V_z(1 - P_z)$. For example, at $V_z = 3$ m^3 and $P_z = 0.8$, the volume V_{st} is equal to: $3 \times 0.2 = 0.6 \text{ m}^3$. The number of equivalent scrap pieces in the zone $n = 0.6/0.196 \times 10^{-3} = 3061$, and the mass of all the pieces M is equal to: $3061 \times 1.55 \times 10^{-3} = 4.7$ т.

In calculations of the equivalent scrap melting rate in a FAF the method of portion charging has been used. As per this method, separate portions are instantly charged into the charging zone with a time space between these portions equal to the time of complete melting down of each portion. Such a method is more convenient for calculations than that of continuous scrap charging. It allows to clearly see the correlation between the following values: the mass of each portion, the time of melting a portion depending on the porosity of the zone at the moment of charging, the number of portions, and the time required for melting the entire mass of scrap. In case of continuous scrap charging the zone porosity depends on a number of conditions and so the calculation becomes less definite.

It should be kept in mind that the scrap melting time calculated in accordance with the portion charging method is slightly higher than that in case of the continuous charging method. Thus, the calculation of FAF's productivity given below is done with a certain "reserve".

At a specified volume of the charging zone an increase in the charging rate decreases the zone porosity and increases the scrap melting time. In order to accelerate melting it is necessary to increase the porosity of the zone due to an increase in its volume. Since the bath depth in a furnace of fixed capacity is also given, the zone volume can be increased only due to an increase in the charging device dimensions which are limited for reasons of design. In the course of the calculations, such a maximal charging rate, at which dimensions of the charging device seem to be acceptable, was determined on a comparative basis of varied alternatives. If any, such parameters as: dimensions of the shaft, the mass of a scrap portion discharged from the shaft at the same stroke of the pusher and its speed required were taken into consideration. The reduced version of the calculation for a FAF of 185 t capacity is stated below.

Initial data required for the calculation of hourly productivity and other performances of the FAF are cited in Table 6.1.[2]

6.2.1.1 Determining of Melting Time of Individual Scrap Pieces in FAF

A cold sample of diameter $d = 25$ mm at a temperature of liquid metal contained in the crucible of induction furnace $t_L = 1650$ °C was melted in 37.5 s. In case of preheating the same sample (or a piece of scrap) up to a temperature of $t_S = 800$ °C its melting time τ_p shortened to 21 s i.e., by 1.78 times [5]. The same reduction in τ_p at $t_S = 800$ °C follows from the analytical formula (3.18), Sect. 3.4.3. This indicates the thoroughness of experimentation by the authors of the work [5] and the high accuracy of the results obtained by them.

---

[2] A number of data in Table 6.1 were obtained as a result of calculations which are not cited here.

Table 6.1 Initial data for calculations

Data group	Data
1. Fuel arc furnace	Capacity 185 t. Tapping weight 130 t Hot heel weigh 55 т
2. Shaft	Height on scrap 11 m. Average cross-section 4.6×4.0 m
3. Pusher	Pusher stroke 5 m. Speed of working stroke 10 m/min Speed of idle stroke 18 m/min Minimal cycle time 43 s
4. Equivalent scrap	Weight 143 t. Volume density $\rho = 0.7$ т/м³ Ratio of surface area F to mass M: F/M $= 20$ m²/t Diameter of piece d $= 25$ mm
5. Duration of periods of the heat	Power-off time 5.5 min. Heating metal after scrap melting down 4.5 min
6. Enthalpies	Scrap at 800 °C: $E_{800} = 145$. Liquid metal at 1570 °C: $E_{1570} = 375$ Liquid metal at 1640 °C: $E_{1640} = 395$ kWh/t
7. Energy efficiency coefficients	Combustion products in shaft $\eta_{bur} = 0.7$. Heat energy of electric arc $\eta_{arc} = 0.87$. Efficiency coefficient of second circuit $\eta_{sec.cir} = 0.92$ Electricity $\eta_{EL} = 0.92 \times 0.87 = 0.80$
8. Input of heat from varied sources	Thermal capacity of natural gas: 10.3 kW/m³. Oxidation of carbon, iron, and its alloys in the bath during melting period $E_{ox.m} = 100$ kWh/t of metal. Oxidation of scrap in shaft $E_{ox.s} = 30$ kWh/t
9. Oxygen injectors	Oxygen density 1.43 kg/m³ (s.t.p.). Oxygen pressure 12 bar (abs.) Pressure of combustible mixture 1.2 bar (abs.) Coefficient of injection U $= 1.7$
10. Certain physical values	Densities: liquid metal 7.0 t/m³, solid metal 7.9 t/m³

These experiments in the induction furnace have been carried out at a temperature of $t_L = 1650$ °C. In the FAF bath maximum value of the t_L during the scrap melting period amounts to 1580 °C. As per formula (3.18), when decreasing t_L from 1650 to 1580 °C the melting time of the sample will increase from 21.0 to 50.4 s.

The value of $\tau_p = 50.4$ s as well as that of $\tau_p = 21.0$ s was obtained when forced stirring of liquid metal in the crucible was absent. In the bath of an electric arc furnace, the liquid metal is intensively stirred by oxygen jets, CO bubbles, and by circulation and pulsation stirring as well [6].

In the FAF, additional measures are offered for the scrap melting rate in the charging zone to increase. For this purpose the roof tuyeres are used for submerged oxygen blowing directed to this zone. In the work [7], heat transfer coefficients from liquid metal to samples melted in induction furnace were determined with and without use of argon blowing of liquid metal. At a liquid metal temperature of 1580 °C the blowing allowed increasing the coefficient α from 12 to 19.5 kW/(m² °C), i.e., by 1.6 times. The melting time τ_p, which is inversely proportional to the α, formula (3.18), must shorten by a factor of 1.6 as well. Because of lack of some other direct data let us assume the value of 1.6 as an index of reduction in τ_p with use of oxygen blowing. Indirect data indicate that a greater reduction in the τ_p is possible due to an improvement of blowing means. For example, replacement of manipulator with consumable pipes in one of EAFs by KT-tuyeres immersed into the slag increased the rate of melting pellets 2.1-fold [8]. Assuming the index of reduction in the τ_p equal to 1.6, let us find that the time of melting a sample, considering not only preheating up to 800 °C but also an effect of improved oxygen blowing on the FAF, amounts to: $50.4/1.6 = 31.5$ s. Let us assume this value of τ_p as the melting time for individual pieces of equivalent scrap as well.

6.2.1.2 Determining of Melting Time of the Entire Scrap and Hourly Productivity of FAF

Multiple pieces in each portion of scrap are melted in the charging zone considerably slower than individual pieces are. In this zone, multiple scrap pieces form a quite closely spaced three-dimensional lattice which has certain hydraulic resistance. This resistance reduces the speed of the metal flows between the pieces. In addition, individual pieces contact to each other which decreases their total surface flowed by the liquid metal. Slowing down of melting, when the porosity of the zone P_z is reduced, is demonstrated by the curve in Fig. 3.7, Chap. 3 [5].[3] Only at $P_z \geq 96$ % the pieces are melted practically independently and the melting rate of the multiple pieces remains equal to approximately that of the individual piece, Fig. 3.7, Chap. 3. However, such a value of P_z cannot be implemented in practice because of too large volumes of the charging zone. From this point of view, scrap charging evenly distributed throughout the area of the furnace bath would be ideal.

With such charging the melting time of 143 tons of scrap in the furnace of 185 t capacity would be equal to about 6 min, Chap. 3, Sect. 3.4.3. The correction of this value may be maximum +20 %. However, even with such a correction the obtained result allows to draw a principal important conclusion. An insufficient scrap melting rate in the furnaces with flat bath is a bottleneck which limits their competitiveness in comparison with EAFs just because the scrap is continuously charged into the liquid metal zone of too small volume. An increase in this volume in real limits can drastically accelerate melting of scrap.

Preliminary calculations of alternatives have shown that in the furnace under consideration the mass of a scrap portion charged M_{st} can be assumed equal to 7.3 t. At such a value of M_{st} all design and mode parameters which determine the scrap melting time and hourly furnace productivity keep within rational limits,

---

[3] Let us recall that in calculations as per the portion scrap charging method the value of P_Z is determined by the mass of scrap portion at the moment of its charging.

well agree to each other and can be realized in practice. Let us determine these parameters using Table 6.1.

If a stroke of the pusher is 5 m and a thickness of the discharged scrap layer amounts to 0.65 m, then the pusher 3.2 m in width fits for the portion mass of 7.3 t. The plate of the charging device as well as the charging zone must have the same width. Let us assume that a plate projection into the furnace freeboard beyond the bottom bank lining bounds is equal to 1.45 m and consider that scrap pieces fall from the end of the plate rather an arc of a circle than vertically. The latter increases a length of the charging zone by 0.15 m i.e., up to 1.6 m. A depth of the metal bath of the 185-t furnace on the average in the course of melting amounts to 0.9 m. Hence it follows that a volume of the scrap charging zone V_z is equal to: $3.2 \times 1.6 \times 0.9 = 4.6 \text{ m}^3$.

The scrap portion mass of 7.3 t is corresponded the value of $V_{st} = 7.3/7.9 = 0.924 \text{ m}^3$ and $P = 1 - 0.924/4.6 = 0.80$. At $P = 80 \%$ the K_τ coefficient is equal to 1.43, Fig. 3.7, Chap. 3. In this case, the melting time of the portion will amount to: $31.5 \times 1.43 = 45$ s (0.75 min), which corresponds to maximum speed of the pusher operation, Table 6.1.

Let us determine scrap charging rate S, the number of portions n, and scrap melting time τ_s. $S = 7.3/0.75 = 9.7$ t/min. Let us notice that maximum scrap charging rate achieved with a 420-t conveyor Consteel EAF amounts to 9 m/min [9]. The number of portions $n = 143/7.3 \cong 20$; and $\tau_s = 143/9.7 = 15$ min.

The total duration of periods of power-off furnace operation and heating a metal amounts to 10 min, Table 6.1. At $\tau_s = 15$ min tap-to-tap time will be equal to 25 min (0.417 h) and hourly productivity of FAF amounts to 312 t/h. Similar hourly productivity is reached only in the biggest EAFs of 300 tons capacity or more.

6.2.2 Electric and Heat Powers, Consumption of Energy Carriers and Electrodes

Let us determine a transformer power during scrap melting and metal heating periods

6.2.2.1 The Scrap Melting Period

During this period the electric arcs must introduce into the metal the heat quantity equal to: $E_{arc} = E_{1570} - E_{800} - E_{ox.m}$, Table 6.1. $E_{arc} = 375 - 145 - 100 = 130$ kWh/t of scrap or $130 \times 143 = 18{,}590$ kWh. At $\eta_{EL} = 0.80$, electrical energy consumption during the melting period amounts to $18{,}590/0.80 = 23{,}237$ kWh. With the period duration equal to 15 min the transformer power will amount to: $23{,}237 \times 10^{-3}/0.25 = 92.9$ MW or considering $\cos\varphi = 0.79$ it amounts to: $92.9/0.79 = 117$ MVA.

6.2.2.2 The Metal Heating Period

Electrical energy consumption during the heating period $E_h = (E_{1640} - E_{1570}) \times 185/0.80$; $E_h = (395 - 375)185/0.8 = 4625$ kWh. With the period duration of 4.5 min (0.075 h), the transformer power will be: $4625 \times 10^{-3}/0.075 = 61.7$ MW or $61.7/0.79 = 78$ MVA. In accordance with the melting period the transformer of 120 MVA is required.

6.2.2.3 Determining the Power of Burner Devices

The heat quantity introduced into the shaft with the burner devices is: $E_{bur} = [(E_{800} - E_{ox.s})/0.7]143$; $E_{bur} = [(145 - 30)/0.7] \times 143 = 23{,}493$ kWh. The devices operation time is 15 min (0.25 h). The power of the devices is: $23{,}493 \times 10^{-3}/0.25 = 94.0$ MW. The total number of devices is three. The power each of them is 31.3 MW.

Electrical energy consumption is: $(23{,}237 + 4625)/130 = 214$ kWh/t.

Natural gas flow rate is: $23{,}493/10.3 \times 130 = 17.5$ m^3/t; or $17.5 \times 130/0.25 = 9100$ m^3/h, or 3000 m^3/h per one burner device.

Oxygen flow rate is: $17.5 \times 2.1 = 36.7$ m^3/t or $36.7 \times 1.43 = 52.5$ kg/t; $9100 \times 2.1 = 19{,}100$ m^3/h or 6300 m^3/h per one burner device.

Electrode consumption is 0.7 kg/t due to shortening tap-to-tap time.

6.2.3 Aerodynamical Parameters of Burner Devices and Shaft

Let us determine the flow rate of gases sucked into the burner devices.

As per data on a finger shaft furnace let us assume that a temperature of gases at the burner devise inlet t_{gas} is equal to about 500 °C [10]. The mass flow rate of these gases is: $G_{gas} = U \cdot G_{O_2}$; U is the coefficient of injection; $U = 1.7$, Table 6.1. $G_{gas} = 1.7 \times 52.5 = 89.2$ kg/t; or $89.2 \times 130/0.25 = 46{,}384$ kg/h. Based on the composition of sucked gases we find their density $\rho = 1.19$ kg/m^3 and the hourly volume flow rate $V_{gas} = 46{,}384/1.19 = 39 \times 10^3$ m^3(s.t.p.)/h.

Let us determine flow rate of gases traveling through the scrap layer located in the lower section of the shaft.

This flow rate is a sum of V_{gas} and a flow rate of combustion products which form when combusting natural gas with oxygen V_{bur}. $V_{bur} = 9100 \times 3.1 = 28 \times 10^3$ m^3 (s.t.p.)/h. Overall flow rate of gases travelling through the scrap amounts to: $(39 + 28) \times 10^3 = 67 \times 10^3$ m^3/h. This flow rate exceeds that of off-gases used for scrap preheating by on the average about 2.0–2.5 times. Therefore, the velocity of gases increases, uniformity of their distribution through the cross section of the shaft improves, and intensity of heat transfer from the gases to the scrap increases as well which contributes to an increase in both the η_{bur} coefficient and scrap preheating temperature.

Table 6.2 Basic
performances of FAF

Performance	Value
Capacity, t	185
Tapping weight, t	130
Tap-to-tap time, min	25
Hourly Productivity, t/h	312
Transformer power, MVA	120
Power of burner devices, MW	94
Electrical energy consumption, kWh/t	214
Natural gas flow rate, m^3/t	17.5
Scrap preheating temperature, °C	800

Calculated performances of the FAF are given in Table 6.2. As follows from this table, the key energy feature of the fuel arc furnace is ultrahigh heat power equal to electrical power. It is this feature in combination with all the innovations required for its realization that provides eventually such performances as productivity and electrical energy consumption which are unachievable for modern EAFs of similar, relatively small capacity. The energy features of a new steelmaking aggregate are emphasized by its heat balance as well. During the scrap melting period, the natural gas introduces the same heat quantity that electricity does. In the modern EAFs, the ratio of natural gas to electrical energy is one to ten.

6.3 Economy and Environment

6.3.1 Economy of Replacement of Electrical Energy with Fuel

The economic expediency of replacement of electrical energy with natural gas is usually determined based on discharge coefficients and on these energy carriers' prices considering some additional factors. The results of such calculation are mainly determined by the ratio of prices on electrical energy and gas which can strongly vary over time and from one country to another. Thus, for instance, in the USA in the period from 2000 to 2005, the price of electricity has barely changed, whereas the price of natural gas has approximately doubled. The cost of 1 kWh of electricity has practically equaled that of natural gas.

Lately, owing to new methods of commercial development of a shale gas deposit, the prices on gas reduced sharply, especially in North America, and they keep going down. Such a situation creates a very favorable economic prospect for utilization of steelmaking aggregates similar to FAFs.

In these new circumstances, there is no need to confirm an obvious economic efficiency of the FAF by specific calculations determining it in terms of money. Such calculations are too often needed to correct. The calculation of efficiency comparing changes in consumption of energy carriers in physical terms is of great interest. The results of these calculations do not depend on price fluctuations. The changes in consumption of energy carriers should be considered rather on the scale of the two interdependent branches of the industry, i.e., of electric steelmaking and of production of electrical energy in the thermal power stations (TPS) than in individual metallurgical plants. Such an approach to the analysis of economic efficiency of FAFs seems to be quite reasonable since in the majority of the countries, in Russia and in the USA in particular, most of the electrical energy is produced in heat power stations (TPS) utilizing as a fuel not coal only, but natural gas as well. About 20 % of electric power in the USA are produced by the gas TPS. In Russia this figure totals approximately 40 %. The combustion of coal releases into the atmosphere considerably larger amount of CO_2, as well as sulfur oxides and other harmful emissions. As a result, substituting gas with coal in TPS would require quite significant additional environmental protection costs. Environment regularities become much stricter. Therefore, one can expect increasing the number of gas TPS especially due to the drop in gas prices.

Let us compare the effectiveness of the use of natural gas for production of electrical energy which later is used in FAF, with the effectiveness of the use of gas directly in the furnace. The overall efficiency coefficient of primary energy of natural gas in the chain of TPS—FAF, η_{TPS}^{FAF}, is determined by the expression:

$$\eta_{TPS}^{FAF} = \eta_{TPS} \times \eta_{EL.GR} \times \eta_{EL} \tag{6.1}$$

η_{EL}—the coefficient of electrical energy efficiency in the furnace

$\eta_{EL.GR}$—the coefficient of efficiency of electrical power grids taken with consideration for all energy losses due to voltage transformations

η_{TPS}—the coefficient of efficiency of the TPS operating on gas

Assuming, in accordance with the current data, $\eta_{EL} = 0.8$, $\eta_{EL.GR} = 0.92$ and $\eta_{TPS} = 0.41$, and using with the expression (6.1), we will obtain $\eta_{TPS}^{FAF} = 0.30$. Therefore, with the use of natural gas in TPS, only approximately 30 % of chemical energy of gas used in TPS is ultimately transferred to the metal bath of FAF heated by the electric arcs. For direct heating of scrap in the furnace shaft the gas energy efficiency coefficient η_{BUR}, is equal to 0.7. Therefore, replacement EAFs with such aggregates as FAFs must lead to in the savings on the overall natural gas consumption in TPS and metallurgical plants expressed in physical terms m^3/t of steel. The calculations of these savings must incorporate the energy consumption for production of oxygen. Let us carry out this calculation using the data from Table 6.2.

In comparison with EAF, the natural gas consumption in FAF increases by $17.5 - 5 = 12.5$ m^3/t, and the oxygen consumption by $12.5 \times 2.1 = 26.2$ m^3/t. The electrical energy consumption for production of oxygen in the modern oxygen-compressor stations is approximately 0.55 kWh/m^3 of O_2. Production of additional 26.2 m^3/t of oxygen requires $26.2 \times 0.55 \cong 14.4$ kWh of electrical energy per 1 t of steel. With an increase in the gas consumption by 12.5 m^3/t, the reduction in the electrical energy consumption in FAF without taking into account the electrical energy consumption for production of oxygen, will be equal to: 370 − 214 = 156 kWh/t. Electrical energy consumption of 14.4 kWh/t for production of oxygen decreases electrical energy savings from 156 to: 156 − 14.4 = 141.6 kWh/t.

In order to produce in TPS and transfer to FAF the 141.6 kWh/t of electrical energy it would be necessary to consume the gas with the heat of combustion equal to 10.3 kWh/m^3 in the amount of 141.6/(0.41 × 0.92 × 10.3) = 36.4 m^3. Thus, per each 1 m^3 of gas additionally consumed in the FAF, on the thermal power station more than 36.4/12.5 = 2.9 m^3 of gas is saved. The absolute savings in the system TPS—FAF amounts to: 2.9 − 1.0 = 1.9 m^3 that is an evidence of an unquestionably high energy efficiency of replacement of electrical energy with gas. If we express these savings in terms of money, then they vary directly proportional to gas prices changing.

The realization of highly productive fuel arc steelmelting aggregates FAFs would allow to implement yet another option of efficient replacement of electrical energy with fuel. We mean the option of combining FAF and oxygen converters at integrated metallurgical plants. Such plants, as a rule, have unutilized resources of coke gas which heat of combustion is equal to 4.1–5.3 kWh/m^3. This gas can be used in FAF in a mixture with natural gas.

The use of fuel arc aggregates can give to the integrated plant yet another additional advantage. It will allow reducing the share of hot metal in the metal-charge processed at the plant due to the use in the FAF of 100 % of the light, lowest-cost scrap. Usually, the share of large and expensive scrap in the charge of the oxygen converters is 25 %. If one fourth of steel at the plant will be produced in the FAF, then the share of scrap in the entire metal-charge processed at the plant will increase to 44 %. This can substantially reduce total production costs at the same output.

6.3.2 Environment

In an FAF as well as in EAFs the problem of dioxins is solved by means of mixing gases leaving the shaft with high-temperature process gases evolved in the furnace freeboard. For this purpose, the gases leaving the shaft are directed to a post-combustion chamber. In the capacity of such a chamber, dust chambers existing in the majority of furnaces for cleaning from large fractions of dust can be used. Off-gases containing CO in a large amount deliver from the freeboard to the same chamber via the gas duct which is an extension of the roof elbow. Carbon oxide is post-combusted with air infiltrated into the gas duct increasing a temperature of gases in the chamber up to the level required for the complete decomposition of dioxins. The CO post-combustion in the furnace freeboard is very inefficient under EAF conditions. In the FAF, CO post-combustion is not only useless but also harmful as far as energy potential of the process gases must be completely consumed for decomposition of dioxins only. The mode of furnace operation with a closed slag door satisfies this requirement.

Aerodynamics of these gas flows has to provide their existing in the chamber during a quite long period of time. This is one of conditions for the complete decomposition of dioxins as well. Atomized water is injected into the gas duct located behind the chamber and leading to the bag filters. This ensures required rapid cooling of gases and eliminates a possibility of recombination of dioxins.

Because of an increase in the natural gas flow rate in the FAF in comparison with that in EAFs the CO_2 emissions into the atmosphere increase as well. However, in the TPS—FAF system these emissions decrease about twofold and in the system with TPS using coal—more than threefold.

6.4 Advantages of Fuel Arc Furnaces

Let us list main advantages of the FAF in comparison with the best EAFs which have identical characteristics such as: capacity of 185 t, tapping weight of 130 t, and transformer power of 120 MVA. These advantages are achieved due to scrap preheating up to a temperature of 800 °C, new means of oxygen blowing, and furnace operation with the flat bath:

- Increased hourly productivity by 36 %, from 230 to 312 t/h
- Reduced electrical energy consumption by a factor of 1.8
- Reduced electrode consumption by 0.4 kg/t
- Reduced total costs on energy-carriers
- Increased yield by 1.5 %
- Reduced max noise level from 120 to 95 dB
- Reduced requirements to power grids and costs for electrical equipment
- Increased reliability and durability of roof and sidewall panels threefold
- Reduced volume of gases requiring cleaning by 30 %
- Reduced CO_2 emissions into the atmosphere two-threefold in TPS—FAF system

At present, there are the COSS-type shaft furnaces with the pushers in a number of countries over the world. In order to equip each of them with the burner devices and oxygen tuyeres described above relatively low investments will be required. Therefore, the FAF project can be implemented in the nearest future. The information given in the book denotes a necessity of concentration of efforts in this regard.

References

1. Memoli F, Guzzon M, Giavani C (2011) The evolution of preheating and importance of hot heel in supersized Consteel® systems. In: AISTech conference, Indianapolis, USA
2. Rummler K, Tunaboylu A, Ertas D (2011) Scrap preheating and continuous charging system for EAF meltshop. MPT Int 5:32–36

3. Abel M, Dorndorf M, Hein M et al (2011) Highly productive electric steelmaking at extra low conversion costs. MPT Int 3:92–96
4. Toulouevski Y, Preisman M, Larry C (2007) A self cleaning water-cooled pressure probe implemented under the EAF roof. MPT Int 2:34–35
5. Li J, Brooks GA, Provatas N (2004) Phase-field modeling of steel scrap melting in a liquid steel bath, vol 1. In: AIS Tech conference, pp 833–843
6. Toulouevski Y, Zinurov I (2013) Innovation in Electric Arc Furnaces: scientific basis for selection, 2nd edn. In: Revised and supplementary, Springer, Berlin
7. Fleisher AG, Kuzmin AL (1982) Effect of temperature of the melt on heat transfer to the surface of an immersed melting body. Izwestiya VUZov. Ferr Metall 4:40–43
8. Malek AO (2004) DC EAF with DRI feeding rates through multipoint injection. MPT Int 2:58–67
9. Mukhopaahyay A, Coughlan R, Ometto M et al (2012) An advanced EAF optimization suite for 420-t jumbo DC furnace at Tokyo Sreel using Danieli technology. In: Proceedings 10th European electric steelmaking conference, Graz, Austria
10. Manfred X, Fuchs G, Auer W (1999) Electric arc furnace technology beyond the year 2000. MPT Int 1:56–63